Science, Technology and Medicine in Modern History

General Editor: **John V. Pickstone**, Centre for the History of Science, Technology and Medicine, University of Manchester, England (www.man.ac.uk/CHSTM)

One purpose of historical writing is to illuminate the present. At the start of the third millennium, science, technology and medicine are enormously important, yet their development is little studied.

The reasons for this failure are as obvious as they are regrettable. Education in many countries, not least in Britain, draws deep divisions between the sciences and the humanities. Men and women who have been trained in science have too often been trained away from history, or from any sustained reflection on how societies work. Those educated in historical or social studies have usually learned so little of science that they remain thereafter suspicious, overawed, or both.

Such a diagnosis is by no means novel, nor is it particularly original to suggest that good historical studies of science may be peculiarly important for understanding our present. Indeed this series could be seen as extending research undertaken over the last half-century. But much of that work has treated science, technology and medicine separately; this series aims to draw them together, partly because the three activities have become ever more intertwined. This breadth of focus and the stress on the relationships of knowledge and practice are particularly appropriate in a series which will concentrate on modern history and on industrial societies. Furthermore, while much of the existing historical scholarship is on American topics, this series aims to be international, encouraging studies on European material. The intention is to present science, technology and medicine as aspects of modern culture, analysing their economic, social and political aspects, but not neglecting the expert content which tends to distance them from other aspects of history. The books will investigate the uses and consequences of technical knowledge, and how it was shaped within particular economic, social and political structures.

Such analyses should contribute to discussions of present dilemmas and to assessments of policy. 'Science' no longer appears to us as a triumphant agent of Enlightenment, breaking the shackles of tradition, enabling command over nature. But neither is it to be seen as merely oppressive and dangerous. Judgement requires information and careful analysis, just as intelligent policy-making requires a community of discourse between men and women trained in technical specialities and those who are not.

This series is intended to supply analysis and to stimulate debate. Opinions will vary between authors; we claim only that the books are based on searching historical study of topics which are important, not least because they cut across conventional academic boundaries. They should appeal not just to historians, nor just to scientists, engineers and doctors, but to all who share the view that science, technology and medicine are far too important to be left out of history.

Titles include:

Roberta E. Bivins
ACUPUNCTURE, EXPERTISE AND CROSS-CULTURAL MEDICINE

Roger Cooter
SURGERY AND SOCIETY IN PEACE AND WAR
Orthopaedics and the Organization of Modern Medicine, 1880–1948

David Edgerton
ENGLAND AND THE AEROPLANE
An Essay on a Militant and Technological Nation

Jean-Paul Gaudillière and Ilana Löwy (*editors*)
THE INVISIBLE INDUSTRIALIST
Manufacture and the Construction of Scientific Knowledge

Thomas Schlich
SURGERY, SCIENCE AND INDUSTRY
A Revolution in Fracture Care, 1950s–1990s

Eve Seguin (*editor*)
INFECTIOUS PROCESSES
Knowledge, Discourse and the Politics of Prions

Crosbie Smith and Jon Agar (*editors*)
MAKING SPACE FOR SCIENCE
Territorial Themes in the Shaping of Knowledge

Science, Technology and Medicine in Modern History
Series Standing Order ISBN 0–333–71492–X hardcover
Series Standing Order ISBN 0–333–80340–X paperback
(*outside North America only*)

You can receive future titles in this series as they are published by placing a standing order. Please contact your bookseller or, in case of difficulty, write to us at the address below with your name and address, the title of the series and one of the ISBNs quoted above.

Customer Services Department, Macmillan Distribution Ltd, Houndmills, Basingstoke, Hampshire RG21 6XS, England

Infectious Processes

Knowledge, Discourse and the Politics of Prions

Edited by

Eve Seguin

palgrave
macmillan

First published 2004 by
PALGRAVE MACMILLAN
Houndmills, Basingstoke, Hampshire RG21 6XS and
175 Fifth Avenue, New York, N.Y. 10010
Companies and representatives throughout the world

PALGRAVE MACMILLAN is the global academic imprint of the Palgrave Macmillan division of St. Martin's Press, LLC and of Palgrave Macmillan Ltd. Macmillan® is a registered trademark in the United States, United Kingdom and other countries. Palgrave is registered trademark in the European Union and other countries.

ISBN 1–4039–3223–9 hardback

This book is printed on paper suitable for recycling and made from fully managed and sustained forest sources.

A catalogue record for this book is available from the British Library.

Library of Congress Cataloging-in-Publication Data
Infectious processes : knowledge, discourse, and the politics of
prions / edited by Eve Seguin.
p. cm.
Includes bibliographical references and index.
ISBN 1–4039–3223–9 (cloth)
1. Prion diseases – History. 2. Prions – History. I. Seguin, Eve, 1963–

QR201.P737I535 2004
614.5′98—dc22 2004043624

10 9 8 7 6 5 4 3 2 1
13 12 11 10 09 08 07 06 05 04

Printed and bound in Great Britain by
Antony Rowe Ltd, Chippenham and Eastbourne

Contents

List of Figures

List of Tables

Acknowledgements

We are grateful to the following for permission to reproduce copyright material:

American Association for the Advancement of Science for Figure 1 of Chapter 1 taken from *Science* 252: 1520.

Macmillan Magazines Limited for Figure 2 of Chapter 1 taken from *Nature* 215: 1044.

John Wiley & Sons Limited for Figure 3 of Chapter 1 taken from R. Brown (chairman): Novel infectious agents and the central nervous system. *Ciba Foundation Symposium* 135: 212.

American Chemical Society for Figure 4 of Chapter 4 taken from *Biochemistry* 31(49): 12278.

Elsevier for Figures 2, 14, 15, 16 of Chapter 4, respectively taken from *Virus Research* 35: 121, *C.R. Biologies* 325: 42 (twice) and *Structure* 91: front cover.

National Academy of Sciences for Figures 3, 5, 6, 10, 11, 14, 15 of Chapter 4, respectively taken from *PNAS USA* 89(22): 10943, *PNAS USA* 91(15): 7141 (twice), *PNAS USA* 97(19): 10088 (twice), *PNAS USA* 94(25): 13456 and *PNAS USA* 98(4): 1462 (twice).

Nature Publishing Group for Figures 12 and 13 of Chapter 4, respectively taken from *Nature Structural Biology* 8: 924 and *Nature* 382: 181.

Scientific American for Figure 1 of Chapter 4, taken from *Scientific American* 251: 49.

Springer for Figures 7, 8, 9 of Chapter 4 taken from *Curr. Top. Microbiol. Immunol.* 207: 57–8.

Prof. Stanley Prusiner for Figures 1, 7, 8, 9 of Chapter 4, respectively taken from *Scientific American* 251: 49 and *Curr. Top. Microbiol. Immunol.* 207: 57–8 (three figures).

Prof. Kurt Wüthrich for Figures 12 and 13 of Chapter 4, respectively taken from *Nature Structural Biology* 8: 924 and *Nature* 382: 181.

We are also very grateful to our commissioning editor, Luciana O'Flaherty, to our series editor, John Pickstone, and to an anonymous reviewer.

Notes on the Contributors

Hanne Andersen is an associate professor in the Department of Medical Philosophy and Clinical Theory at the University of Copenhagen, Denmark. She previously worked as curator at the Danish National Museum for the History of Science and Medicine. Her research focuses on the nature and processes of conceptual change in science, specifically on how incommensurability develops, and on its implications for the realism debate. Among her recent publications are the monograph *On Kuhn* (Wadsworth, 2001), and articles in such journals as *Philosophy of Science, Studies in History and Philosophy of Science* and *Erkenntnis*.

Brigitte Chamak worked as a neurobiologist for ten years. She then decided to study the history and sociology of science and did a PhD on the emergence of cognitive science in France. Her book *Le Groupe des Dix* was published by Editions du Rocher in 1997. She is currently working as a sociologist of science in the Cesames (Centre de recherche Psychotropes, Santé Mentale) at the Université Paris 5, France. After doing research on the prion hypothesis, she is now studying the controversies generated by different representations of autism. Her articles have been published in such journals as *La Pensée, Alliage, Intellectica, La Revue d'Histoire des Sciences* and *Social Studies of Science*.

Kerstin Dressel is a sociologist and is currently the scientific secretary of the German TSE Research Platform at the University of Munich, Germany. She is also a member of the board of the South German Institute of Empirical Social Research in Munich. She has conducted several social scientific studies on various aspects of the BSE case for the EU, WHO, OECD and the German Parliament. She has published several papers on the issue and her book *BSE – the New Dimension of Uncertainty* was published by Edition Sigma in 2002. Her research interests encompass the science–policy interface, social and cultural studies of science, scientific knowledge and ignorance, and risk and policy research.

Eric Francoeur is a sociologist of science who specializes in the use of models in chemistry. He has worked at the Max Planck Institute for the History of Science in Berlin and in the Centre de Sociologie de l'Innovation at the Ecole Nationale Supérieure des Mines de Paris. He is currently lecturer at the Ecole de technologie supérieure de Montréal,

Canada. He has published articles in such journals as *Hyle: an International Journal for the Philosophy of Chemistry* and *Social Studies of Science.*

Maj-Britt Juhl Poulsen is a research adviser in the Health Care Sector in Frederiksborg County, Denmark. After working as a researcher in the field of molecular biology at Statens Serum Institut in Copenhagen, she gained a PhD in Science and Technology Studies. Her research focuses on competition and cooperation in biomedical research, on the development of biomedical research, and on research integrity. Her recent publications are found in *International Journal of Technology Management* and *Proceedings of the First ORI Research Conference on Research Integrity.*

Ki-Heung Kim recently completed his PhD thesis on the history of scrapie research in the Science Studies Unit at the University of Edinburgh. He is now a research fellow in the Wellcome Trust Centre for the History of Medicine at University College London, UK, where he is doing a comparative study on the history of research on neurological diseases in the UK and USA.

Jérôme Segal trained as an engineer and did a PhD in the history of science on the scientific concept of information. He has worked on the history of molecular biology at the Max Planck Institute for the History of Science in Berlin. He is now an assistant professor at the Institut Universitaire de Formation des Maîtres in Paris and a researcher at the Centre Cavailles of the Ecole Normale Supérieure de Paris, France. He has made several contributions on cybernetics, information theory and the history of molecular biology to edited volumes, and has published articles in such journals as *Annals of Science.* His book *Le zéro et le un* was published by Syllepse in 2003.

Eve Seguin is a political scientist and a researcher in the Department of History at the University of Aberdeen, UK. Her research interests encompass public controversies on technoscientific issues, science studies, the links between science and politics, discourse analysis, and the use of expertise in policy-making. In case studies such as *in vitro* fertilization or BSE, her work specifically addresses the political function of scientific discourse. She has published a number of articles in such journals as *Theory, Culture and Society, Revue française de science politique, Discourse and Society, Revue française d'administration publique, Science as Culture, Langage et société, Science and Public Policy, Studies in History and Philosophy of Science.*

List of Abbreviations

ABRO	Animal Breeding Research Organization (UK)
AFSSA	Agence Française de Sécurité Sanitaire des Aliments (France)
AHS	Army Health Service (SSA: Service de santé des armées) (France)
AIDS	acquired immune deficiency syndrome
AFRC	Agricultural and Food Research Council (UK)
ARC	Agricultural Research Council (UK)
BSE	bovine spongiform encephalopathy
cDNA	complementary DNA
CEA	Centre d'Energie Atomique (France)
CJD	Creutzfeldt–Jakob disease
CNEVA	Centre National d'Etudes Vétérinaires et Alimentaires (France)
CNRS	Centre National de la Recherche Scientifique (France)
CNS	central nervous system
CWD	chronic wasting disease
DH	Department of Health (UK)
DMA	Délégation Ministérielle pour l'Armement (France)
DNA	deoxyribonucleic acid
DRET	Direction des Recherches et Etudes Techniques (France)
EU	European Union
GH	growth hormone
GIS	Groupement d'intérêt scientifique (France)
GSS	Gerstmann–Sträussler–Scheinker syndrome
HIV	human immunodeficiency virus
HSP	heat-shock proteins
IAH	Institute of Animal Health (UK)
IBR	Institute of Basic Research (USA)
INRA	Institut National de la Recherche Agronomique (France)
INSERM	Institut National de la Santé et de la Recherche Médicale (France)
IRAD	Institute for Research on Animal Diseases (UK)
MAFF	Ministry of Agriculture, Fisheries and Food (UK)
MIT	Massachussetts Institute of Technology (USA)
mRNA	messenger RNA
MRC	Medical Research Council (UK)

NMR	nuclear magnetic resonance
NPU	Neuropathogenesis Unit (UK)
PCR	polymerase chain reaction
PNS	peripheral nervous system
PrP	prion protein
PrP^{C}	cellular (normal) isoform of PrP
PrP^{Sc}	scrapie (pathological) isoform of PrP
PVR	poliovirus receptor
RKI	Robert-Koch Institute (Germany)
RML	Rocky Mountain Laboratory (USA)
RNA	ribonucleic acid
SAFs	scrapie associated fibrils
SEs	spongiform encephalopathies
SEAC	Spongiform Encephalopathies Advisory Committee (UK)
SSI	sociology of scientific ignorance
SSK	sociology of scientific knowledge
TME	transmissible mink encephalopathy
TSEs	transmissible spongiform encephalopathies
UC	University of California (USA)
UCSF	University of California in San Francisco (USA)
vCJD	new variant of Creutzfeldt–Jakob disease
WOS	Web of Science

Introduction: Prions?

Eve Seguin

On 20 March 1996, the Secretary of State for Health, Stephen Dorrell, announced in the British Parliament that the ten cases of a new variant of Creutzfeldt–Jakob disease (vCJD) discovered in the UK had probably been caused by the consumption of beef infected with bovine spongiform encephalopathy (BSE). This announcement marked a radical shift in the discourse of the government and provoked one of the most notable public health crises that Europe had faced in the twentieth century. For weeks and months, what now appeared as the transmission of BSE to humans became a major concern of the media, featuring on broadcast programmes and on the front page of all newspapers.

One striking feature of this new episode of the BSE saga was the recurrent appearance of 'prions', thought to be the infectious agents of the two diseases. The tentative explanation was that the bovine prion had jumped the species barrier to infect humans. Of course, prions as the causative agents of transmissible spongiform encephalopathies (TSEs) postulated by American scientist Stanley Prusiner, had already featured in the media. However, the March 1996 statement led to a massive escalation of media reports and allowed prions to make their official entrance in the public domain.

Since March 1996 the BSE crisis has been extensively scrutinized by academics, and a number of articles and books have been published. Notably, scholars tend to tackle BSE in much the same way as the public inquiry set up by the UK government in December 1997. That is, the social-scientific literature focuses on the politics of BSE and is mainly concerned with explaining the mishandling of the crisis by government officials and politicians. When this literature deals with the scientific dimension of BSE, it is mostly in terms of scientific uncertainty. Many authors insist on the lack of knowledge about spongiform encephalopathies during the BSE epidemic. If prions come up at all in these works, they are never the focus of analysis.

This is surprising since prions display a number of interesting and paradoxical features. Let me enumerate just a few. First, prions are enigmatic agents usually described as 'infectious proteins'. Second, it has been claimed that the prion concept is akin to a scientific revolution in biology. Third, Stanley Prusiner was awarded the 1997 Nobel Prize for physiology or medicine for his discovery of prions, 'a new biological principle or infection'. Finally, if the prion hypothesis, which holds that TSEs are caused by the abnormal, infectious, form of the prion protein, is now widely accepted by the scientific community, many researchers are still disputing it.

Strangely enough, the prion saga has also been largely ignored by historians and sociologists of science. In 1999, Martha Keyes published in a journal devoted to the history of biological sciences a two-part article which discussed the prion challenge to the Central Dogma of molecular biology. Was it the signal that prions were now being recognized as scholarly interesting creatures? Unfortunately not. Though BSE was a popular topic at the joint 4S/EASST conference held in Vienna in 2000, not a single paper on prions was delivered. Such was the situation when I started to work on the links between the BSE crisis and the prion hypothesis.

Faced with this void, I decided to systematically search for other historians and sociologists who might be working on prions. I was lucky enough to find a few scholars whose work had not yet been published. I also approached academics who were not working on prions at that time but whose expertise was likely to make an important contribution to this topic. Most of these researchers were isolated from one another and to bring them together I organized a panel on prions at the 2001 annual conference of the 4S held in Cambridge, USA. Four of the eight contributors to this volume delivered a paper and the panel convinced me that we collectively held a body of knowledge that should be organized and communicated.

The present volume is thus intended to fill (part of) the gap in our sociohistorical understanding of prions. Chapter 1 outlines the early history of the protein-only hypothesis in scrapie research and explains why Prusiner successfully promoted it when his predecessors had failed in the 1960s. Chapter 2 describes the controversy between supporters and opponents of the prion hypothesis and argues that the former gained the upper hand due to their style of scientific practice. Chapter 3 compares the development of prion work to Kuhn's model of scientific change and argues that Prusiner's prion hypothesis is a new paradigm under construction. Chapter 4 emphasizes the growing importance of computer modelling in the work of Prusiner and other scientists, and shows how visualization contributed to solidify the epistemic status of prions. Chapter 5 shows that prion research has greatly benefited from the BSE crisis and emphasizes the role played by actors external to the scientific community in the promotion of prion work. Chapter 6 provides a quantitative analysis of the circulation of prion discourse in scientific literature and argues that it was decisive in the recognition of the prion research programme.

We hope this volume will shed some analytical light on one of the most important episodes in late twentieth-century science.

1

The Early History of the Protein-only Hypothesis: Scientific Change and Multidisciplinary Research

Maj-Britt Juhl Poulsen and Hanne Andersen

Introduction

In 1997, the American neurologist and biochemist Stanley B. Prusiner received the Nobel Prize in medicine for his discovery of 'prions' – a new biological principle of infection. Preceding this discovery lies a complicated history of the research on a number of neurodegenerative diseases, including the sheep disease scrapie. During the 1960s, research on scrapie revealed that the infectious agent had very unusual characteristics, and a variety of hypotheses regarding the principle of scrapie infection were advanced. However, not until the 1980s did a single hypothesis, Prusiner's prion hypothesis – which is basically a protein-only hypothesis[1] – succeed in attracting the attention of the majority of the scientific community. Interestingly, protein-only hypotheses had been advanced as early as the mid-1960s but without receiving any notable support from other scientists than those who had advanced them.

The acceptance of the hypothesis that proteins can act as infectious agents, despite their lack of hereditary material, has been a revolution in contemporary biological thinking, and the early history of the hypothesis illuminates the complicated process that precedes such major scientific changes. Based on a detailed analysis of this history we shall examine what was required to obtain acceptance of the hypothesis.[2] We shall focus on the recognition of anomalies and the detailed elaboration of alternative theories, as well as the communication difficulties inherent in multidisciplinary research in order to give a plausible explanation of why it took 20 years for the protein-only hypothesis to gain acceptance.

Scientific change

Scientific change has been an important issue in the philosophy of science since the 1960s. Initiated by Kuhn's 1962 classical *The Structure of Scientific Revolutions* and the subsequent work of scholars like Lakatos (1970), Toulmin (1972), and Laudan (1977), scientific change has become a central ingredient in the understanding of the growth of scientific knowledge. However, it has been a matter of recurrent disputes whether scientific change is a rare or a frequent occurrence, and whether all scientific changes imply a wholesale replacement of ideas or if they may imply only minor changes.[3] Among the central issues concerning scientific change have been the questions of how and why scientific change happens, how a new theory gains acceptance in the scientific community, and what characterizes the researchers that advance new, radical ideas.

In most analyses of scientific change it has been assumed that the research area – whether this was described by a Kuhnian paradigm, a Lakatosian research programme, Laudanian guiding assumptions, or in other terms – could be characterized by rather well-defined research problems, methods, and values. New insights may therefore be gained from analysing a contemporary multidisciplinary research field like the research on transmissible neurodegenerative diseases, a field that draws on a multitude of different methods, theories, and scientific values. In this chapter we shall focus on some aspects of scientific change that are very sensitive to the question of multidisciplinarity. That is, how the judgement of anomalies and the reception of alternative hypotheses may be dependent on disciplinary perspectives, and how this affects the possibility for researchers working interdisciplinarily to initiate scientific change.

Judgement of anomalies

It holds both in monodisciplinary as well as in multidisciplinary research that anomalies are not all equally severe. Some discrepancy between theoretical predictions and experimental findings can always be found but without questioning the foundation of normal science research. For an anomaly to be a severe anomaly that leads to questioning the accepted tools and understandings, it must have some special importance. The number and age of anomalies as well as their demonstrated resistance to solution are some of the factors that may influence how far a received scientific view is questioned (see, for example, Laudan, 1977: 36ff.). But also the cognitive importance of an anomaly is essential in assessing whether it questions the accepted theory (see, for example, Kuhn, 1962: 82; Laudan, 1977: 37f.). Thus, an anomaly is likely to be severe if it calls some very fundamental generalizations into question, or if it calls into question achievements that have a particular practical importance (Kuhn, 1962: 82). However, different scientists may judge anomalies differently and therefore disagree on when an old theory seems untenable. Such differences in the judgement of anomalies may be dependent on, for example, differences in background knowledge (Barker et al., 2002). Since scientists working in multidisciplinary fields may often draw on different background knowledge, differences in the judgement of anomalies may therefore be outspoken in such fields.

Generation of alternative theories

The severity of an anomaly depends not only on its cognitive importance and resistance to solution. As noted by Laudan, the importance of an anomaly for a theory depends also on the competitive state of play between that theory and its competitors (Laudan, 1977: 38). This ties the assessment of anomalies inextricably to the generation of alternative hypotheses.

Much recent work on anomalies focuses explicitly on the cognitive processes of anomaly resolution, that is, the kind of scientific reasoning involved in localizing an anomaly and generating new hypotheses that can account for it (for example Darden, 1992, 1998). In some cases anomalies may be explained away without requiring theory change. This can be done by claiming that experimental error occurred, or by claiming simply that the case is not a normal instance, but a monstrous one; what Darden calls a 'monster anomaly' (Darden, 1992: 258f; 1998: 142). Other anomalies require a change. However, after the theory change the anomaly ceases to be an anomaly and can now instead be viewed as an instance of the normal. This kind of anomaly Darden calls 'model anomalies' because they serve as models of normal types of processes that are commonly found (Darden, 1992: 259; 1998: 143).

Whether an alternative hypothesis advanced in response to a model anomaly is accepted by the scientific community depends on several things. As described in general terms by, for example, Kuhn and Laudan, to accept a new theory requires that it solves the problems which called the previous theory into question. Further it should display a quantitative precision strikingly better than its predecessor. Finally, it should predict new and hitherto unexpected phenomena. Other values can be included as well, such as, for example, the internal consistency of the competing theories, their clarity, their extendibility and fruitfulness, and their relations to other accepted theories (for example Darden, 1992: 262). Again, most work on theory assessment has focused on single disciplines with relatively well-defined values. However, in a multidisciplinary field scientists with backgrounds in different disciplines may vary considerably not only in their assessment of an anomaly, but also in when the anomaly can be considered solved, and in what counts as, for instance, clarity or fruitfulness.

Initiating change in a multidisciplinary field

One of the implications of the early accounts of scientific change was that the more committed scientists are to prior practice and the more they know about the solved problems and the expectations for future research, the more is at stake in a change of theory and the more reluctant will they be to suggest alternatives. In contrast, people who are young in their field and therefore less committed to the tradition than their senior colleagues may suggest alternatives more easily. This led Kuhn and others to suggest that scientific change often is initiated by younger scientists or scientists who are in other ways relatively new to their field (for example Kuhn, 1962: 151; similarly Feyerabend, 1970: 203; Cantor, 1975: 195).[4] Adapted to multidisciplinary research this suggests that scientists might suggest changes more easily in areas different from their original training. However, due to differences in scientific background scientists in a multidisciplinary research field may adopt very different standards as to what research contributions should look like, which methods to apply, what constitutes a good argument, and so on.[5] Hence, although scientists who cross disciplinary boundaries might be more likely to suggest radical changes, their research may not be well received within the discipline for which the change is suggested.

The history of scrapie research as an instance of scientific change in a multidisciplinary field

In investigating these aspects of scientific change in multidisciplinary research, the history of scrapie research offers an outstanding opportunity to examine a contemporary case involving scientists from very diverse areas such as veterinary medicine, medicine, biochemistry, molecular biology, radiobiology, theoretical chemistry and biophysics. This case offers the opportunity to study the crossing of disciplinary boundaries, the judgement of anomalies, and the response to controversial suggestions by researchers with different disciplinary backgrounds.

The background history of scrapie research

Scrapie in sheep is a fatal, degenerative disorder of the central nervous system (CNS) and has been known since before the middle of the eighteenth century (M'Gowan, 1914). Initially some scientists believed that the disease was transmitted from parent to offspring and M'Gowan suggested that scrapie was caused by a heavy infection of the progeny with a protozoan parasite (sarcosporidium) (M'Gowan, 1914). This was questioned by other investigators and the possibility of a virus was tentatively mentioned (for instance M'Fadyean, 1918; Stockman, 1926). A major step in understanding this disease was the successful transmission of scrapie from infected to healthy sheep in 1936 (Cuillé and Chelle, 1936). The transmission of scrapie from sheep to goats proved that it was possible to jump the species barrier and the infectious agent was suggested to be a 'filterable' virus (Cuillé and Chelle, 1939). The suggestion that the scrapie agent was a virus was made on the basis of filtration experiments, yet its conventional character was soon questioned. Through the 1940s and the early 1950s the first reports on the unconventional character of the scrapie agent started to emerge (for example Gordon, 1946; Greig, 1950), and in 1954 scrapie was grouped together with several other animal diseases in a new category of virus infections termed 'slow infections' (Sigurdsson, 1954). This new category was suggested in contrast to the known categories 'acute infections' and 'chronic infections'. One characteristic of slow infections was a very long initial period of latency lasting from several months to several years, but by creating a special class for this abnormal phenomenon, the anomaly was explained away. Other puzzling characteristics of the disease were the failure to display any immune response (Chandler, 1959), that, apparently, it could arise spontaneously (Pattison and Millson, 1960, 1961a), and that, apparently it could be both an infective and genetic disease (for example Parry, 1962). Furthermore, different forms of the disease with different clinical symptoms (drowsy or scratchy) were described, and a species barrier between different species was recognized (for example Pattison and Millson, 1960, 1961b; Pattison, 1966a, b). However, classified into the abnormal class of 'slow infections', these anomalous characteristics did not present themselves as serious anomalies.

During the late 1950s, scrapie was linked to human diseases for the first time. The resemblance of the human diseases kuru and Creutzfeldt–Jakob disease (CJD) was recognized in 1959 (Klatzo et al., 1959), and the same year scrapie and kuru were connected (Hadlow,

1959). However, it was not until the mid and late 1960s that the transmissibility of kuru and CJD to chimpanzees was established (for example Gajdusek et al., 1966; Gibbs et al., 1968; Gibbs and Gajdusek, 1969).[6]

For a long time research on scrapie progressed only slowly, partly because it proved very difficult to purify the scrapie agent (for example Hunter and Millson, 1964; Stamp, 1967; Pattison, 1970), partly because the only way to measure the presence of the infectious agent was by detecting its ability to induce the disease in animals (for example Greig, 1950; Stamp, 1967; Pattison, 1970). A major step forward was accomplished when the scrapie agent was passed from sheep into mice, which have a relatively short incubation period for the disease (Chandler, 1961). This made many new experimental studies possible that could not be performed on sheep or goats. With mice as guinea pigs, scrapie became the most amenable of the diseases to study experimentally.

Anomalies in the search for an infectious agent and the first 'non-nucleic acid' hypotheses

During the 1960s reports on the unusual properties of the infectious agent accumulated. These properties included the finding that the agent was quite small (Hunter and Millson, 1964, Alper et al., 1966) and highly resistant to heat (for example Stamp, 1962; Hunter and Millson, 1964), to formalin (for example Greig, 1950; Pattison, 1965), and to ultraviolet light (Alper et al., 1966, 1967). Several of these chemical and physical treatments were known to inactivate conventional viruses and bacteria, and it was suggested that the scrapie agent could not be fitted into the pattern of any known infective particle (for example Pattison, 1965; Alper et al., 1966; Gibbons and Hunter, 1967; Adams and Field, 1968).

In trying to explain these unusual properties, several researchers developed various hypotheses both on the chemical structure of the infectious agent, on the mechanism for replication, and on the pathogenesis. The hypotheses on the nature of the infectious agent included, among others, different suggestions concerning unconventional viruses, the suggestion that it was a polysaccharide part of membrane, or that it consisted only of protein.

Tikvah Alper from the Hammersmith Hospital in London was one of the researchers who advanced alternative hypotheses in response to the anomalies which the virus hypothesis was facing. Working within the fields of radiobiology and radiation chemistry, she counted among her main research interests bacterial and viral radio sensitivity and the radiation target sizes of biological macromolecules. When experiments made by her and her collaborators showed both that the size of the infectious agent was exceptionally small and that it was not inactivated by irradiation with ultraviolet light of a wavelength known to inactivate viruses generally (Alper et al., 1966, 1967), she had found two anomalies to the virus theory which both fell within her main areas of expertise. On the basis of their first experiments, Alper and collaborators suggested that 'the agent may be able to increase in quantity without itself containing nucleic acid' (Alper et al., 1966: 283). Alper and collaborators had found an anomaly which from their perspective as radiobiologists was so severe that it called for radical changes. By suggesting that infectious agents exist which do not contain nucleic acid, they were taking the first step to transform the anomaly into a model anomaly that could serve as a model for a new type of infectious agent.

However, they also realized that 'the responsibility of the nucleic acids for replication and genetic control is so firmly established ... that this suggestion has not been generally acceptable' (Alper et al., 1967: 764). To strengthen their argument that the scrapie agent had to be distinguished from viruses, forming instead a new class of infectious agents, they continued their inactivation experiments with ultraviolet light. After a year they confirmed their findings in a paper in *Nature*: 'The results confirm our previous conclusion that scrapie is most unlikely to depend on a nucleic acid moiety for its replicative ability' (Alper et al., 1967: 764). In their publications they kept mentioning that they had considered the virus theory to be very firmly established and that, therefore, they had to strengthen their anomalous results as much as possible if they were to hope for acceptance of a completely new category of infectious agent. For example, in a 1978 publication they still emphasized that 'nucleic acid was at that time so firmly established as the only biological compound with "self-replicating" properties that it was clearly desirable to find some means of testing our inference' (Alper et al., 1978: 504).

Another opponent to the virus hypothesis came from a different field of research: veterinary medicine. Based at the Agricultural Research Council's field station in Compton, I. H. Pattison had been working on the pathology of scrapie since the late 1940s. During the 1950s, he had followed the work of another Compton-based researcher, D. R. Wilson, on the unconventional characteristics of the scrapie agent (Pattison, 1988). During the 1960s, Pattison's own research showed further anomalous characteristics of the agent, most notably its surprisingly high resistance to formalin. Based on their experiments on the physico-chemical properties of the agent, Pattison and his collaborator Katherine Jones concluded that 'the agent has unusual physico-chemical properties, closely similar to those of encephalitogenic factor, *which is believed to be a basic protein*' (Pattison and Jones, 1967: 7, italics added). They therefore advanced the idea 'that the transmissible agent of scrapie may be, or may be associated with, a small basic protein' (Pattison and Jones, 1967: 2). They further hypothesized that it was 'possible to visualize a replicating mechanism not hitherto recognized in mammalian cells' although 'only further work can establish the mechanism of replication of the scrapie-transmissible agent' (Pattison and Jones, 1967: 8). Hence, like Alper and collaborators, Pattison and Jones attempted to turn the anomalous scrapie agent into a model anomaly that would form a new type of infectious agent. However, as their latter remark suggests, the suggestion that the infectious agent did not contain nucleic acid entailed major explanatory problems within the existing theoretical framework in molecular biology, including that of replication.

The non-nucleic acid hypotheses contradicted the Central Dogma of molecular biology, that is, the conception that nucleic acid is the hereditary material and that information transfer from protein to protein is impossible. However, analysing Francis Crick's original version of the Central Dogma and Watson's later and narrower interpretation of it reveals that Pattison's hypothesis about a replicating protein contradicts only Watson's interpretation but not Crick's original version (Keyes, 1999a,b). Crick's original version from 1958 stated that *sequential information* transfer is possible from nucleic acid to nucleic acid, or from nucleic acid to protein, and forbade the transfer from protein to protein or from protein back to nucleic acid – but it did not directly address the concept of replication. In 1965, Watson advanced a much narrower version of the Central Dogma in which he equated the flow of sequential information with the *production* of nucleic acids and proteins. According to this interpretation, which was the theoretical framework within which molecular biologists approached the puzzle of the scrapie agent, the protein-only hypothesis was therefore a controversial suggestion: an infectious replicating protein was a completely new concept in molecular biology. Still in the 1980s when Prusiner renamed the scrapie agent 'prion', this hypothesis of a replicating, infectious agent consisting of no more than protein was denounced by many scientists as 'heretical' on the grounds that it contradicted the Central Dogma of molecular biology.

The first formulations of an alternative protein-only theory

The major problem facing the protein-only hypothesis, the explanation of the self-replication of proteins, attracted the attention of a scientist who had not previously been involved in scrapie research: J. S. Griffith. In 1961, he had received an ScD in theoretical chemistry, and had been employed first as a professor of mathematics in Manchester, later as a professor of applied mathematics in the Department of Mathematics at Bedford College in London. His previous research had primarily been within quantum chemistry.

In a paper published in the prestigious journal *Nature* in September 1967, Griffith set out to show that there are at least three distinct classes of mechanisms that would allow proteins to replicate and that, consequently, 'there is no reason to fear that the existence of a protein agent would cause the whole theoretical structure of molecular biology to come tumbling down' (Griffith, 1967: 1043).

The first possible mechanism of protein self-replication he considered is that of a protein acting as an inducer:

> It is possible that, through a chance mutation, a gene G may arise which is switched off in all cells of a particular animal. If this mutation is not selectively disadvantageous, it will simply persist. Suppose, furthermore, that G codes for a protein S which acts as an inducer for G. Then S could act as an infective agent because it is normally never present but, if it is introduced, it will reproduce itself. (Griffith, 1967: 1043)

Griffith then examined which properties of scrapie such a model would predict. On this model, it would not be possible to predict the long period of latency, but it would not be surprising either. However, the model could predict a number of other facts about scrapie. First, further mutations of G might explain the difference in sensitivity of sheep. Second, since the protein S might induce a gene G′ similar to the gene G but in a different animal, the model might also explain why other animals than sheep are susceptible to scrapie. Third, the occurrence of two distinct forms of the disease could be explained by two genes G_1 and G_2 with corresponding products S_1 and S_2. Fourth, since G may not be repressed with absolute efficiency, the model might also explain how scrapie can arise spontaneously.

The second mechanism of self-replication considered by Griffith was inspired by a mechanical analogue to self-replication which had been described ten years earlier in a letter in *Nature* by the medical doctor and psychiatrist L. S. Penrose and his son Roger who was trained in mathematics and physics. According to Penrose and Penrose, the self-reproducing properties of nucleic acid did not necessarily depend upon its highly complex structure. Instead, they presented a device exemplified by pieces of plywood or vulcanite which despite their simple characters possessed the property of being able to replicate (Penrose and Penrose, 1957). By analogy, Griffith suggested a model building on the polymerizing of different conformations of a protein, stating his argument by means of reaction schemes and free energy changes.

The model built on a series of assumptions regarding the energy balance of various reactions involving the stable and normal cellular protein structure α' which may change its conformation into another conformation α. Further, the model built on the polymerizing of the two different conformations. Thus, Griffith assumed that the production of α' and the production of polymers of α are energetically privileged, although some of the latter may not take place directly. Given a certain minimum relation between the energies released in the various polymerizing processes, a polymer may act as a catalyst for a complicated process in which it may reproduce. Thus, Griffith imagined that an existing α_2 or α_3 may act as a template that enables the α' to fold into α while joining onto the template. Thus, if only α' is a normal cellular constituent whereas α and α_2 are not, α_2 can, by its catalytic role in the process, act as an infectious agent that reproduces itself. Such a model would be able to explain the existence of two clinically distinct forms of scrapie by the formation of mixed polymers from two possibly isozymic forms of α', that is, chemically distinct but functionally identical forms of the enzyme. Further, the model could explain the spontaneous appearance of scrapie by a spontaneous conversion of α' into α which immediately combines into α_2.

The third mechanism considered by Griffith was that the protein acts as an antigen (A) which stimulates the production of an antibody (A′). Usually, the antibody is different from the antigen, but if the two are identical, the process would enable the protein to replicate itself.

Griffith noted that there was experimental evidence that the disease is not antigenic, thus excluding the third model. However, two models for protein replication still remained, and Griffith ended his paper concluding:

> If it belongs to one of the first two classes, then it is a protein or a set of proteins which the animal is genetically equipped to make, but

> which it either does not normally make or does not make in that form. It may be passed between animals but be actually a different protein in different species. Finally, in either case, there is the possibility of spontaneous appearance of the disease in previously healthy animals. (Griffith, 1967: 1044)

Thus, Griffith had provided a solution to the biggest problem facing a protein-only hypothesis: that of explaining replication in the absence of hereditary material. This result might have been a breakthrough for the protein-only hypothesis – but, for reasons that will be explained later, it was largely ignored by the scientific community.

The interlude

Both Alper and Pattison continued their work on the scrapie agent. During the 1960s and 1970s, Alper and co-workers presented further evidence against the scrapie agent's dependence for replication on intrinsic nucleic acid. They cited the protein-only hypotheses advanced by Pattison and by Griffith, and even though they did not regard the radiobiological evidence to be in conflict with the protein-only hypotheses, they favoured the hypothesis that the determining factor in scrapie is a rearrangement in the sugar or oligosaccharide residues attached to the cell membrane (Gibbons and Hunter, 1967). Hence, in 1978 they stated of the scrapie agent that 'a lipid fraction is an important component and to that extent provides additional support for the "membrane hypothesis"' (Alper et al., 1978: 503).

Pattison cited his 1967 paper five times during the 1960s and 1970s, but although he repeated his original hypothesis that the infectious agent might be, or might be associated with, a small basic protein, he also considered another hypothesis that did not involve the self-replication of the agent. Thus, in 1970 he argued that 'the progressive increase in amount of scrapie agent in the tissues of inoculated animals may be due to unmasking [that is, the scrapie agent is present in an inhibited form in normal tissue and in a released form in scrapie tissue] rather than to multiplication, as hitherto assumed' (Pattison, 1970: 673).

Pattison did not cite Griffith in any of these five papers. Griffith only cited his own 1967 paper twice, namely in two papers in *Journal of Theoretical Biology* (Griffith, 1968a,b), which mainly contain mathematical discussions of equations describing the regulation of mRNA and protein synthesis but do not discuss the scrapie agent as such. Further, he only cited his 1967 paper briefly, arguing that the discussions may 'subsequently [be] used as a basis for explanatory hypotheses about various biological problems' (Griffith, 1968a: 202). Although he did add a single limiting condition to one of the proposed models of self-replication in one of his 1968 papers,[7] there is no trace in his work that he pursued his own suggestions about the protein-only model of scrapie, at least not in a biochemical sense showing that an *actual* reaction of the proposed type could be carried out.

However, not all researchers found it necessary to depart so radically from the conventional virus theory as Alper and Pattison had done in order to explain the many exceptional properties of the agent. As Pattison summarized the situation in 1970:

> Although the characteristics of the scrapie agent...are generally accepted as factual, their significance has been interpreted by different workers in different ways. Broadly speaking, scrapie investigators can now be divided into two categories. One category includes those who maintain that the agent is a virus, and, by inference, that it contains nucleic acid. For example, a conventional virus (e.g. Eklund, Kennedy and Hadlow, 1967; Gajdusek and Gibbs, 1968), or a virus with certain unusual features (Stamp, 1967), or a virus with a polysaccharide coat and small nucleic acid core (Adams and Caspary, 1967) or a provirus that is produced *in vivo* and multiplies to produce disease (Parry, 1962). The second category includes those workers who believe that the agent does not contain nucleic acid and that it is not a virus. For example, a replicating polysaccharide (Field, 1966), part of the nucleoprotein complex (Pattison and Jones, 1967), a rearrangement in sugar or oligosaccharide residues in cell membrane (Gibbons and Hunter, 1967), an entity associated with cellular plasma membrane (Hunter, 1969), or a linkage substance present in normal tissues (Adams and Field, 1968). The present confusion in scrapie research is clearly reflected in these widely varying interpretations of the nature of the scrapie agent. Those who support a virus etiology cannot easily explain the physico-chemical properties of the agent, especially the extreme resistance to ultraviolet irradiation. Those who do not favour a virus hypothesis cannot explain animal-to-animal passage, that presumably signifies replication of the agent. (Pattison, 1970: 676–7)

Hence, by 1970 the scientific community was split into two groups: one that preferred the established theory and treated the anomalous characteristics of the scrapie agent as monster anomalies that could be explained away by categorizing it in the abnormal class of slow viruses. The other group had started treating the anomalous characteristics as model anomalies and now struggled to model a new class of infectious agents from their results.

During the 1970s, the debate about the constituents of the scrapie agent continued. Different variants of the virus theory were still defended by some researchers (for example Narang, 1974; Rohwer and Gajdusek, 1980), and even Prusiner, who later became famous for the prion hypothesis, started out supporting the virus theory (Masiarz et al., 1980). But new hypotheses opposing the virus theory were also advanced, such as, for example, that the scrapie agent was a replicating RNA (a so-called viroid) (Diener, 1972, 1973), or that the agent was a *Spiroplasma*, that is, a group of wall-free prokaryotes (for example Bastian, 1979). During this period, no single hypothesis was able to convince the majority of the scientific community. As Pattison later recalled it:

> I retired in 1976, with the nature of the transmissible agent of scrapie still obscure, and virologists as adamant as ever that theirs was the only possible point of view.
>
> Years passed without progress. Then, out of the fog, a new name emerged. In 1982 S.B. Prusiner put forward the prion hypothesis, postulating that 'novel proteinaceous particles cause scrapie'. (Pattison, 1992: 21)

Watson's stringent interpretation of the Central Dogma prevailed throughout the 1960s (Keyes, 1999a,b). In 1970, Howard Temin and David Baltimore discovered the enzyme reverse transcriptase which is capable of copying RNA into DNA (Baltimore, 1970; Temin and Mizutani, 1970). This clearly violated Watson's stringent interpretation of the Central Dogma, but only with regard to the relation between DNA and RNA. Thus, this did not change the theoretical framework regarding the relation between RNA/DNA and protein and between protein and protein. It was still a controversial suggestion to imagine a self-replicating protein; a suggestion that involved a new class of infectious agents radically different from any known infectious agents.

The prion theory

At the beginning of the 1980s, a protein-only hypothesis was advanced again, now by Stanley Prusiner from the University of California in

San Francisco. Trained as a medical doctor, Prusiner had started working on spongiform encephalopathies (SEs) in 1972 when he lost a patient to CJD (Prusiner, 1995). During the 1970s, Prusiner and collaborators[8] worked on improving the techniques by which the infectious agent of scrapie could be purified (Prusiner, 1995). In 1981, on the basis of their examination of a highly purified preparation of scrapie agent, they were able to advance the hypothesis that the scrapie agent *contains* a protein (Prusiner et al., 1981). However, in 1982 Prusiner went one step further and suggested that the *dominant* characteristics of the agent resembled those of a protein (Prusiner, 1982).[9] He now coined the scrapie agent 'prion' for *pro*teinaceous *in*fectious particle (Prusiner, 1982: 141). In the opening of the paper Prusiner referred to the anomalous characteristics of the scrapie agent as 'enigmatic properties' and the 'mysteries surrounding the scrapie agent' (Prusiner, 1982: 136). These anomalous characteristics had now been used to create a new category of infectious agents that had been given a name, and which Prusiner thought might in the future help to identify the aetiologies of other degenerative diseases such as Alzheimer's senile dementia, multiple sclerosis, Parkinson's disease, or rheumatoid arthritis (Prusiner, 1982: 143).

In the article introducing the new term prion, Prusiner summarized the evidence established by several scientists that a protein is *required* for infectivity. Having established that a protein is required for infection, Prusiner discarded several of the hypotheses on the structures and mechanisms of the scrapie agent which had been advanced, such as the polysaccharide hypothesis and the nucleic acid-only hypothesis. Still emphasizing the unconventional properties of the scrapie agent, Prusiner argued that both his own data and those of other investigators suggested 'two possible models for the scrapie agent: (i) a small nucleic acid surrounded by a tightly packed protein coat, or (ii) a protein devoid of nucleic acid, that is, an infectious protein' (Prusiner, 1982: 141).

Discussing the second possibility, that prions are devoid of nucleic acid, he suggested two alternative models for their replication. One model was that the prion acts as an inducer, and resembled the first model contained in Griffith's 1967 paper on possible models of protein replication. Prusiner described the mechanism of this model as an active transcription of a host gene coding for prion protein, and argued that if cellular genes coding for the scrapie prion do exist, they must be highly regulated, not readily activated, and present in various mammalian cells ranging from mice to monkeys. An occasional activation of such cellular genes might then explain the sporadic occurrence of CJD. The other model was built on the possibility that the prion acts as a template for its own synthesis. However, at this point Prusiner did not provide any details of such a mechanism, but indicated that 'unorthodox mechanisms such as reverse translation or *protein-directed protein synthesis* would allow prions to replicate' (Prusiner, 1982: 142, italics added). Furthermore, Prusiner referred to early studies on crystalline tobacco mosaic virus where no RNA was found, and to the suggestion that the protein of tobacco mosaic virus was autocatalytic.

Later the same year, Prusiner and collaborators identified a protein which proved to be unique to preparations from scrapie-infected brains (Bolton et al., 1982; Prusiner et al., 1982). The protein was designated PrP, for prion protein, and data soon indicated that PrP was a component of the prion (Mckinley et al., 1983). PrP was the first structural molecule within the scrapie agent that was identified. Further, on the basis of their experimental data, the team claimed that it was the *only* major protein contained in the prion, although minor proteins could not be excluded, just as it was impossible to say how many PrP molecules a single prion consisted of.

In an article in *Scientific American* published two years later, Prusiner presented several other models for the production of PrP, one of them resembling Griffith's second model involving different conformations of a protein. Prusiner presented this model in a diagrammatic form, as one of seven hypothetical mechanisms, and ascribed it to the category of mechanisms in which prion proteins are encoded in the genome of the host animal (Prusiner, 1984). Prusiner argued in the legend: 'The requirement of biologically active PrP molecules for the synthesis of new prion particles could also be explained *if PrP catalyzes the conversion of a precursor molecule into PrP*' (Prusiner, 1984: 55, italics added).

During the following six years, Prusiner was the author or co-author of nearly 150 publications about prions and the SEs (Seguin, Chapter 6 this volume). A huge amount of data was accumulated by Prusiner's group as well as other research groups around the world on scrapie, on other SE diseases, and on the PrP protein.[10] By the mid-1980s, the purification and characterization studies of scrapie prions were linked to the discovery that the neurodegenerative formations in scrapie-affected brains are composed of prion protein. Furthermore, the chromosomal gene encoding PrP was identified, and the possibility that PrP might be encoded by a nucleic acid in the scrapie agent was excluded (for example Oesch et al., 1985; Chesebro et al., 1985; Basler et al., 1986; Barry et al., 1986). It was established that PrP is produced both in normal and infected tissue, the normal cellular PrP gene product (designated PrP^{C}), and the scrapie PrP (designated PrP^{Sc}) seemingly having the exact same amino acid sequence (Basler et al., 1986; Prusiner, 1991). This led the Prusiner group to suggest the possibility of variations in protein conformation:

> Our results suggest that there is only a single PrP gene, and its sequence and organization make it unlikely that the different properties of the PrP isoforms can be explained by alterations in the amino acid sequence; thus, it seems more probable that the isoforms arise from posttranslational modifications *or variations in protein conformation.* (Basler et al., 1986, italics added)

Tentative models of how PrP^{Sc} is capable of inducing the disease and directing the production of more of itself were advanced at the Ciba Foundation Symposium in 1987 (Bolton and Bendheim, 1988; Oesch et al., 1988). As one possible solution Prusiner and co-workers advanced the 'direct action' model, which held that 'PrP^{Sc} might act like an enzyme, converting either PrP^{C} or its precursor into PrP^{Sc} which in turn would catalyze further conversion' (Oesch et al., 1988: 211). Other American researchers proposed a 'modified host protein' model in which the specific protein modification may be either covalent or non-covalent, either by acting on the normal protein directly or by affecting one of the steps in protein processing (Bolton and Bendheim, 1988). Thus, both papers introduced the possibility of a direct interaction between the normal and abnormal PrP proteins.

In the following year, the genetic linkage between PrP mutations and GSS in humans was established by the Prusiner group 'providing the best evidence to date that this familial condition is inherited despite also being infectious' (Hsiao et al., 1989: 343). Moreover, experiments indicated that the barrier for transmission between species resides in the amino acid sequence of PrP (for example Scott et al., 1989). Prusiner saw these and several other findings as vindicating his prion hypothesis, and in 1990 he and his collaborators advanced a detailed model for prion replication:

> In attempting to reconcile all of the currently available experimental data on the molecular structure of prions, we propose a model for scrapie prion replication in which existing PrP^{Sc} molecules serve as a template for the posttranslational conversion of PrP^{C} or a precursor into similar PrP^{Sc}. (Prusiner et al., 1990: 674)

From radiation experiments there was some evidence about the minimum size of the putative PrP^{Sc}–PrP^{C} complex, but uncertainty remained as to the chemical details of the conversion process. As Prusiner and collaborators summarized:

> Whether this conversion of PrP^{C} or a precursor into PrP^{Sc} involves the addition or deletion of a chemical group, a tightly bound ligand, or only a conformational change remains to be established. To date, there is evidence for neither a chemical modification nor a ligand that is unique to the PrP^{Sc} isoform. These observations raise the possibility that the difference between PrP^{C} and PrP^{Sc} *is only conformational.* (Prusiner et al., 1990: 681, italics added)

On this latter model, strains or isolates of scrapie prions that breed true are explained by assuming that different stable conformations may act as templates for the conversion of PrP^{C} into each their PrP^{Sc} molecules. Further, this model is consistent with various results showing that PrP can exist in multiple conformational states (for example Hay et al., 1987; Yost et al., 1990). Finally, if mutant PrP^{C} molecules could fold spontaneously into the appropriate conformation for PrP^{Sc} at some relatively low but finite frequency, the model could explain the finding that the GSS syndrome is both a genetic and an infectious disease.

In a 1991 single-authored paper (Prusiner, 1991), Prusiner presented the model in a diagrammatic form substantiating his theory with further experimental data (see Figure 1.1). In this paper Prusiner additionally suggested that 'foldases', chaperones, or other macromolecules feature in the conversion of the PrP^{Sc}–PrP^{C} heterodimer to PrP^{Sc} molecules, and referred to studies of conformational changes in other enzymes to make his hypothesis plausible.

With this detailed model for prion replication, the major problem in creating a new category of infectious agents on the basis of the many anomalous findings on the scrapie agent had been solved. The protein-only hypothesis was still considered controversial, but the work of Prusiner and collaborators had gradually received more and more response from the scientific community and this development continued. Prusiner was awarded the Nobel Prize for medicine in 1997 for his discovery of prions.

Reception of the protein-only hypotheses

The papers by Alper and colleagues (Alper et al., 1966, 1967) proposing that scrapie is most unlikely to depend on a nucleic acid moiety for its replicative ability, the paper by Pattison and Jones (1967) cautiously

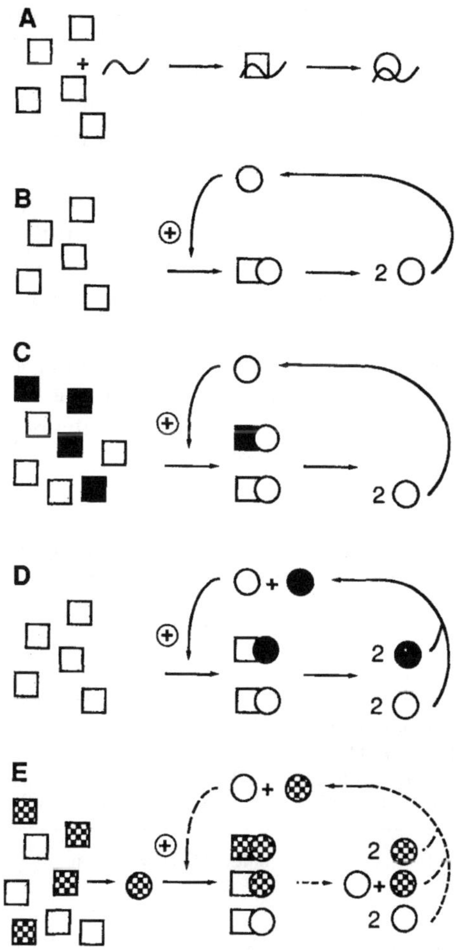

Fig. 4. Some possible mechanisms of prion replication. (**A**) Two-component prion model. Prions contain a putative, as yet unidentified, nucleic acid or other second component (solid, thick wavy line) that binds to PrP^C (squares) and stimulates conversion of PrP^C or a precursor to PrP^{Sc} (circles). (**B**) One-component prion model—prions devoid of nucleic acid. PrP^{Sc} binds to PrP^C forming heterodimers that function as replication intermediates in the synthesis of PrP^{Sc}. Repeated cycles of this process result in an exponential increase in PrP^{Sc}. (**C**) Prion synthesis in transgenic mice (*71*). $HaPrP^{Sc}$ (circles) binds to $HaPrP^C$ (white squares), leading to the synthesis of PrP^{Sc}. Binding to $MoPrP^C$ (black squares) does not produce PrP^{Sc}. Species barrier for scrapie between mice and hamsters represented by $MoPrP^C$-$HaPrP^{Sc}$ heterodimer. (**D**) Scrapie isolates or strains in hamsters or mice. Multiple PrP^{Sc} conformers (circles) bind to PrP^C and constrain the conformational changes that PrP^C undergoes during its conversion into PrP^{Sc}. (**E**) Inherited prion diseases in humans and transgenic mice. Mutant PrP^C molecules (checkered pattern in squares) might initiate the conversion of PrP^C to PrP^{Sc} (or PrP^{CJD}). If infectious prions are produced (dashed lines), then they stimulate the synthesis of more PrP^{CJD} in humans and PrP^{Sc} in experimental animals. Alternatively, prion infectivity is not generated, but the host develops neurologic dysfunction, spongiform degeneration, astrocytic gliosis, and possibly PrP amyloid plaques (*2, 3, 60, 61*).

Figure 1.1 Diagrammatic illustrations of the model of replication proposed by the Prusiner group

Source: Prusiner (1991). (Reprinted with permission from *Science*, 252: 1520. Copyright 1991 American Association for the Advancement of Science.)

advancing a protein-only hypothesis, as well as the paper by Griffith (1967) providing possible mechanisms of protein replication necessary for a protein-only hypothesis to be plausible received only little response from the scientific community during the first two decades after their publication, as shown in Table 1.1. In the paper in which the term 'prion' was introduced, Prusiner briefly listed previous hypotheses on the chemical structure of the scrapie agent. Among these was a replicating protein with a reference to Griffith's paper. Most of the subsequent citations of Griffith's paper mention it only as an early hypothesis of a replicating protein without further discussion of the content.

In contrast, the papers in which Prusiner advanced his protein-only hypothesis, suggested the term 'prion', and made the first tentative suggestion that the prion may act as a template for its own synthesis, received an almost overwhelming interest within a very short period of time, as shown in Table 1.2.

Table 1.1 Citation analysis of the core papers by Alper, Pattison and Griffith

Publication	*Number of citations, SCI 1966–99*						
	1966–69	*1970–74*	*1975–79*	*1980–84*	*1985–89*	*1990–94*	*1995–99*
Alper et al. (1966)	33	35	18	20	40	25	34
Alper et al. (1967)	24	34	11	15	52	59	81
Pattison and Jones (1967)	25	9	3	5	4	12	5
Griffith (1967)	4	6	3	5	13	40	105

Source: Science Citation Index, CD-ROM edition.

Table 1.2 Citation analysis of the core papers advancing the prion theory

Publication	*Number of citations, SCI 1980–99*			
	1982–84	*1985–89*	*1990–94*	*1995–99*
Prusiner (1982)	65	198	216	302
Basler et al. (1986)	–	84	156	110
Oesch et al. (1988)	–	–	30	7
Prusiner et al. (1990)	–	–	122	178
Prusiner (1991)	–	–	230	349

Source: Science Citation Index, CD-ROM edition.

How can this difference between the reaction towards the hypotheses advanced by Alper, Pattison and Griffith on the one hand, and the hypothesis advanced by Prusiner on the other, be explained? We shall show in the following section that both the judgement of anomalies and the generation and acceptance of new hypotheses are important factors in scientific change that are very sensitive to scientists' disciplinary background.[11]

From monster anomaly to model anomaly

As described above in the section 'Scientific change', scientists usually only advance or adopt an alternative hypothesis if severe anomalies have put the reigning theory in a state of crisis. However, scientists may differ in their assessment of such anomalies. Scientists involved in scrapie research came from a variety of fields and may therefore have had very different views on the scrapie anomalies.

For Alper, the anomalous radiobiological findings were at the core of her research interests. She had more than a decade's experience with the inactivation of infectious agents by radiation. Thus, radiobiological findings fell in her area of expertise, and they ran counter to established results on the inactivation of infectious agents. What may for researchers outside the field of radiobiology and radiation chemistry have been merely unconventional characteristics of a virus were for Alper anomalies so severe that the scrapie agent could not just be explained away as an instance of an abnormal class of viruses. For her these anomalous findings had to be turned into model anomalies, and that called for an alternative theory of infection, even if it ran counter to a firmly established theory.

Pattison had been involved in scrapie research for decades and had witnessed the growth of anomalous findings. He had met his own results on the high resistance of the scrapie agent to formalin with 'disbelief', yet, much to his 'amazement' new experiments confirmed it (Pattison, 1988: 663). When adding this 'very considerable' resistance to formalin to all the other unusual findings, Pattison also began to turn these anomalous results into model anomalies that questioned the reigning virus theory.

Hence, several researchers saw the anomalous findings as a challenge to the virus theory. However, treating these findings as model anomalies and instances of the normal implied that a new type of infectious agent had to be modelled on the findings. The first problem was that even among the researchers who questioned the virus theory, there was little agreement on what this new category of infectious agent was. Suggestions included not only a protein but also a replicating polysaccharide, part of the nucleoprotein complex, a rearrangement in sugar or oligosaccharide residues in cell membrane, an entity associated with cellular plasma membrane, or a linkage substance present in normal tissues. Rather than working in consonance on a single theory, most of the opponents of the virus theory followed their own favourite alternative.

The next problem was to develop the hypothesis related to the new infectious agent in such detail that it could solve the problems that had led the virus theory into trouble. The protein-only hypothesis did in fact explain the main experimental findings which were anomalous for the virus theory such as the agent's small size and its resistance to physical and chemical treatments known to inactivate viruses. However, there was a severe problem facing the protein-only hypothesis: that of explaining the self-replication of a protein. If the protein-only hypothesis was to be plausible, this problem had to be solved. Although Alper and her group, as well as Pattison and his collaborators, continued to state that the increase in the amount of scrapie agent in infected tissue seemed to be independent of the intrinsic nucleic acid of the agent, the problem of how this increase could take place had to be solved if their hypotheses were to gain acceptance. Griffith's work was an attempt to provide a model for the self-replication of a protein, but it was not treated as a solution to the problem by those who needed it. Neither Alper nor Pattison pursued the models of self-replication advanced by Griffith, but saw instead their own results as vindications of other hypotheses like the membrane hypothesis or the unmasking hypothesis. Alper only mentioned Griffith's model as one possible solution to self-replication but without supporting it (Haig et al., 1969; Alper et al., 1978), and Pattison never cited Griffith's 1967 paper. Without a solution to the problem of self-replication, it was impossible to establish proteins as a new category of infectious agents.

Insiders and outsiders

On a superficial view one of the models of replication which Griffith suggested was more or less similar to that later proposed by Prusiner.[12] However, although Griffith advanced different models for the self-replication of proteins, these did not appear to the scientific community of scrapie research as *solutions* to the problem. His second model, that of conformational change during polymerizing, was stated as a hypothetical argument that the 'reaction scheme could be realized through the assignment of physically reasonable properties to the protein subunits' (Griffith, 1967: 1044). Such a hypothetical argument showed a *possible* solution, but it did not show that this was also the best solution to the problem.

Since Griffith was a theoretical chemist and biophysicist, it is not surprising that he did not pursue in an experimental, biochemical way, the model he had proposed. But did his model invite such a biochemical pursuit in the first place? The main assumptions of the model were assumptions regarding the energy balance of various reactions involving different conformations of a specific protein. Testing these assumptions empirically would be no easy task – if possible at all at that time.[13] Furthermore, as the work of a theoretical chemist and biophysicist, Griffith's suggestions may not have appealed to the experimentalists dominating the scrapie research field. Griffith's article built its argument on mathematical equations. It drew on the remote field of thermodynamics. It presented no experimental evidence. Indeed, as shown in Figure 1.2, it was not a conventional biochemical or medical paper.

The linguistic and diagrammatic expressions of Griffith's article are very different from those seen in Prusiner's later prion articles. Importantly, Griffith offers no diagrammatic illustration of his model of replication. An indication of how his paper may have been looked upon by scrapie researchers can be found in Hunter's later recollection that

> Griffith proposed in *Nature* a more general proposition that the scrapie agent was a protein. Griffith made some interesting suggestions about the possible self-replication of proteins, but there were at that time no ways of testing for the transmission of information back from protein into nucleic acid to reverse the functional direction of the genetic code. It was really pure speculation rather than hypothesis. (Hunter, 1992: 26)

Not only did Hunter dismiss Griffith's suggestions as 'pure speculation rather than hypothesis' – apparently he even misunderstood (or, at least, severely misrecollected) what Griffith's paper was about, since it had nothing to do with the transmission of information back from protein into nucleic acid! Having forgotten the true but unconventional content of the paper, Hunter recollected it instead along the lines of a conventional biochemical argument. Those aspects of his work in which Griffith had expertise – the theoretical underpinning in terms of thermodynamics – were never really acknowledged, and what was taken into account appeared as mere speculation. Likewise, when the similarity

The Second Way. Self-replication need not involve any very intricate mechanism, provided that suitable components are available[9]. Penrose and Penrose illustrated this with a mechanical example[10] but we shall now see that a logically similar one could be made out of protein.

Given protein sub-units α which can undergo the following reactions

$$\begin{aligned} \alpha_2 + \alpha &\rightarrow \alpha_3 \\ \alpha_3 + \alpha &\rightarrow \alpha_4 \\ \alpha_4 &\rightarrow 2\alpha_2 \end{aligned} \tag{1}$$

it is clear that the net result is that the monomer α gets converted into the dimer α_2. If we suppose also that the reaction

$$\alpha + \alpha \rightarrow \alpha_2 \tag{2}$$

cannot take place directly, it follows that α can only dimerize to α_2 under the catalytic influence of molecules of α_2 which are already present.

Next we show that this reaction scheme could be realized through the assignment of physically reasonable properties to the protein sub-units. Let the reactive sub-unit α be a different conformation of the stable structure, α', say, of the protein. Then we have the equilibrium

$$\alpha' \rightarrow \alpha - \Delta F_1 \tag{3}$$

The free energy change ΔF_1 could easily be so large (for example, > 100 kT) that the mean total number of sub-units in all cells in the animal, which are in the form α, is much smaller than unity. We suppose, then, that all sub-units are in the conformation α'. The reaction

$$\alpha + \alpha \rightarrow \alpha_2 + \Delta F_2 \tag{4}$$

may then be very favourable to the formation of dimers but can never proceed because no molecules in the conformation α are available. In fact if $\Delta F_2 > 2\Delta F_1$, the reaction

$$\alpha' + \alpha' \rightarrow \alpha_2 + \Delta F_2 - 2\Delta F_1 \tag{5}$$

is thermodynamically favourable to α_2 but cannot actually proceed. (I mean by this, strictly, that it will be an enormously long time before even a single molecule of α_2 appears.)

The remaining reactions are

$$\begin{aligned} \alpha_2 + \alpha' &\rightarrow \alpha_3 + \Delta F_3 \\ \alpha_3 + \alpha' &\rightarrow \alpha_4 + \Delta F_4 \\ \alpha_4 &\rightarrow 2\alpha_2 + \Delta F_5 \end{aligned} \tag{6}$$

with $\Delta F_3 > 0$, $\Delta F_4 > 0$, $\Delta F_5 > 0$. The first two reactions of equations (6) involve the combined combination of α' and change of its conformation. That this, unlike reaction (5), should proceed at measurable velocity is quite plausible because α' can fold into its new conformation as it joins onto the "template" α_2 or α_3. Finally, we can combine equations (5) and (6) to show that

$$\Delta F_5 = -2\,\Delta F_1 + \Delta F_2 - \Delta F_3 - \Delta F_4 > 0 \tag{7}$$

which is true if both ΔF_3 and ΔF_4 are small. The first of these conditions requires no justification. The second could be interpreted to mean that the tetramer α_4 is very

Figure 1.2 Griffith's thermodynamical argument for the possibility of protein replication

Source: Griffith (1967) (Reprinted with permission from *Nature*, 215: 1044. Copyright 1967 Macmillan Magazines Limited.)

between his second model and the model proposed by the Prusiner group finally drew attention to his work during the 1990s, it was often referred to as a 'protein-only' hypothesis, with no further details. Thus, the impression his work finally left was mostly the idea of a mechanism for self-replication of proteins, and not the thermodynamical argument.[14]

The field of thermodynamics is a remote field to many experimental researchers engaged in biomedical work, and Griffith's thermodynamical argument may not have been understood or even properly noticed by other researchers engaged in the scrapie field. Still, for these researchers it may have been possible to get an idea of a mechanism for self-replication of proteins by reading Griffith's article, even though they may not have understood the thermodynamical argument.

Whereas Griffith was an outsider to the field of SE research and had problems conveying his results to SE researchers, Prusiner, in constrast, came from the biomedical field. During the 1970s and 1980s, he gradually climbed the academic ladder to become full professor of neurology, virology, and biochemistry at the UCSF and UC, Berkeley.[15] Between 1982 and 1990, he published more than 100 papers in several prestigious neurology, virology and biochemistry journals such as *Nature, Science, Cell, Biochemistry, Proceedings of the National Academy of Sciences of the United States of America, Lancet, New England Journal of Medicine, Journal of Infectious Diseases, Journal of Neurochemistry*, and *Journal of Virology*. Most of these articles are reports of experimental results on the SEs, and advancing his prion hypothesis he did not deviate from this pattern. All his major hypotheses regarding prions were advanced in some of the most prestigious journals such as *Science* and *Cell*. Further, at the early stage of his work on the SEs, Prusiner collaborated directly with established researchers in the field. For example, some of his papers were co-authored with Bill Hadlow, a veterinarian who had worked for decades on scrapie and had described the similarities between kuru and scrapie in a letter to *Lancet* in 1959.

As shown in Figure 1.3, Prusiner's papers on the prion hypothesis were expressed in the ordinary linguistic and diagrammatic way for biomedical papers.

As described in the section 'The prion theory' above, Prusiner continually substantiated his suggestions with experimental results, and pointed to future experiments to be performed. Additionally, he continuously repeated his controversial suggestions about prions and how they replicate, and accumulated data to substantiate them. The outcome of his efforts is a prevalent citation rate both by the Prusiner group itself

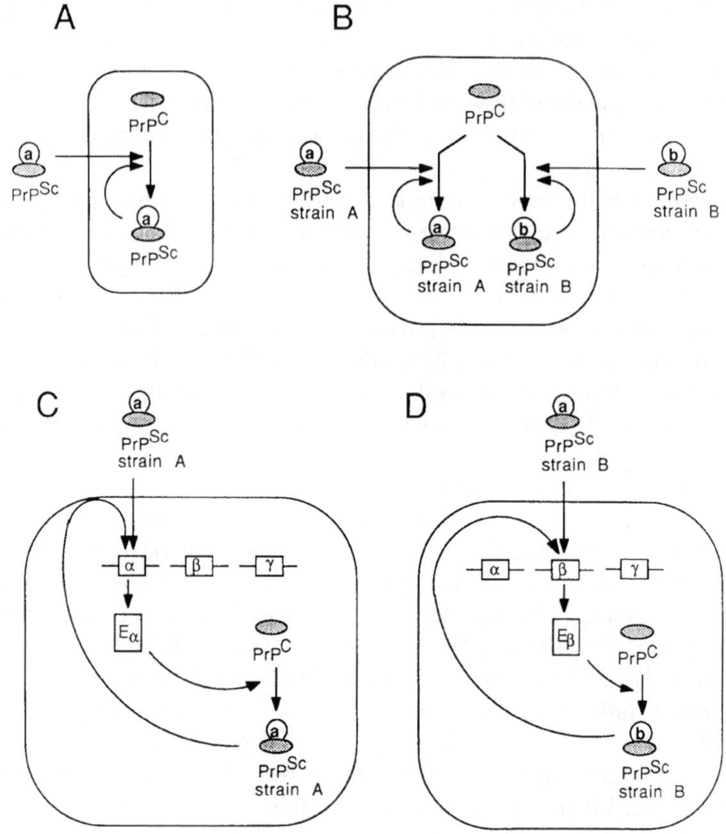

FIG. 1. Models for the replication of scrapie agent without a nucleic acid component ('protein only'). PrP^{Sc} is the infectious agent. (A,B) 'Direct action' model. (A) On infection, PrP^{Sc} converts PrP^{C} (or its precursor) into PrP^{Sc}, which in turn leads to further modification of PrP^{C}. (B) The replication of different strains of PrP^{Sc} is explained by assuming that strain specificity resides in different modifications (a,b) which direct the conversion of PrP^{C} into PrP^{Sc} carrying the same modification as the infecting strain. (C,D) 'Indirect action' model. PrP^{Sc}, strain A, activates a cellular gene α, whose product, E_{α}, converts cellular PrP into PrP^{Sc}. To explain the propagation of different strains of PrP^{Sc} in a single host genotype, one would postulate the existence of a battery of such cellular genes, each encoding a differer.• 'PrP converting enzyme'.

Figure 1.3 Diagrammatic illustrations of the model of replication proposed by the Prusiner group

Source: Oesch et al. (1988). (From R. Brown (chairman): *Novel Infectious Agents and the Central Nervous System. Ciba Foundation Symposium,* 135: 212. Copyright 1988. © John Wiley and Sons Limited. Reproduced with permission.)

Table 1.3 Differences between Griffith the outsider and Prusiner the insider

	Griffith	*Prusiner*
Training and research fields	Theoretical chemistry Applied mathematics	Neurology Virology Biochemistry
Publication style	Linguistic and diagrammatic expressions do not conform to biochemical or medical papers	Ordinary linguistic and diagrammatic biomedical features
Empirical foundation	No empirical evidence and no suggestions of experimental follow-up	Empirical evidence and suggestions of further experimental follow-up

and by other research groups in SE research (Seguin, Chapter 6 this volume).

In short, as captured in Table 1.3, Griffith did not meet the requirements to successfully promote a protein-only hypothesis, whereas Prusiner undoubtedly did.

Conclusion

In conclusion, several criteria had to be fulfilled before the controversial protein-only hypothesis could be recognized and eventually accepted. From the mid-1940s, the virus theory had struggled with anomalies which the protein-only hypothesis could resolve – yet this alone did not suffice, since the protein-only hypothesis faced another problem, that of explaining protein replication.

Radiobiologist Alper and co-workers had taken an initial step to turn anomalies into model anomalies by suggesting that the infectious agent did not contain nucleic acid, but they did not attempt to elaborate the details of an alternative theory that could explain this new category of infectious agents. Veterinarian Pattison and his collaborators also suggested that a new category of infectious agents should be introduced, but they only hypothesized on the characteristics of this new category and did not elaborate on it. A solution to the problem of replication in the absence of nucleic acid had to be formulated to turn the observed anomalies into model anomalies. The theoretical chemist and biophysicist Griffith made just such an attempt, but his alternative hypotheses were not recognized by the SE scientific community as a possible object for their research, including experimental testing. The work of Griffith did not seem to solve the problems which called the virus theory into question, and therefore his proposed models did not gain acceptance in the scientific community.

Plausible explanations of the little attention which the work of Alper, Pattison and Griffith received from the scientific community are:

1. That the early papers were not followed up as vigilantly as would be required to persuade the relevant scientific community;
2. Alper and Pattison stayed experimentally within their own fields and did not see their results as vindicating the same hypothesis;
3. Griffith did not explain himself in an ordinary, biomedical sense; and
4. It is not clear how to pursue the work of Griffith experimentally.

During the 1970s, changes were made in the framework of molecular biology, but these almost only concerned the relation of the nucleic acids, and did not make it any easier to suggest self-replication of proteins. To most researchers a replicating protein was still a heretical suggestion which was not easily embraced. It was only through concerted efforts including biochemical experiments, transmission studies and inactivation studies that the protein-only hypothesis gradually gained acceptance in the scientific community.

In this case the research was directed by the multiple trained neurologist, virologist and biochemist Stanley Prusiner and included a large research network. Prusiner was a specialist in several fields and had the ability to comprehend multiple disciplines and to communicate across disciplinary boundaries. Unlike Alper, Pattison and Griffith, whose work did not develop to a degree that enabled it to cross disciplinary boundaries of the SE community, the work of Prusiner and collaborators had the concerted ability to communicate anomalies and at the same time to present alternative hypotheses in such a way that made his suggestions possible to adopt in the whole community of SE. Hence, it may be that an outsider (in this case Griffith) is more likely to suggest radical changes, but it has to be done in a way that makes it amenable to the insiders (the SE community in 1967). Furthermore, what this case study suggests is that although, in a multidisciplinary field, individual researchers may recognize anomalies within their particular area of expertise and formulate alternative hypotheses, the transformation of the anomaly into a model anomaly and the subsequent development and acceptance of an alternative theory in the whole multidisciplinary field, requires the concerted action of a broad range of expertise.

Acknowledgements

Research for this study was supported by the Danish Research Council for the Humanities. We would also like to thank Peter Barker and Eve Seguin for valuable comments and suggestions on earlier versions of this chapter.

Notes

1. The prion concept started out to encompass a family of hypotheses stating that a protein was required for infectivity. One of these hypotheses asserted that prions are composed entirely of protein (Prusiner, 1999: 79). It is this protein-only hypothesis Prusiner now defends, though broadly defining a prion as a 'proteinaceous infectious particle that lacks nucleic acid' (Prusiner, 1999: 80).
2. It is not our concern in this chapter to discuss who to credit or who was right in this scientific debate. Our concern is solely to analyse the protein-only hypotheses and the reaction of the scientific community to these particular hypotheses.
3. See, for example, McMullin (1992) or Andersen (1996) for arguments that revolutionary change includes 'mini-revolutions' in which continuity prevails, although not exclusively.
4. This thesis should not be conflated with the similar theses that young scientists are more likely to accept a new theory than older scientists (for historical case studies against this view, see Hull et al., 1978 and Messeri, 1988), or that when a new field is opened up to investigation, younger scientists are more likely to be drawn into it than older ones (for a case study in support of this view, see Rappa and Debackere, 1993).
5. It must be noted that these communication problems are different from what Kuhn labelled incommensurability. Whereas incommensurability is a relation between two theories addressing the *same* object domain, the communication problems with which we are here concerned arise between scientists drawing their arguments from *different* object domains.
6. Today several animal diseases including scrapie of sheep and goats and bovine spongiform encephalopathy (BSE) of cattle (in public entitled 'mad cow disease') and several human diseases including kuru, CJD, and Gerstmann–Sträussler–Scheinker syndrome (GSS) have been grouped together as transmissible spongiform encephalopathies (TSEs) or as prion diseases. However, in this chapter, to avoid historiographical confusion as to when the individual diseases were recognized as transmissible, we refer to this group of diseases as SEs (spongiform encephalopathies). All these diseases are fatal since they cause a neurodegenerative process in which the brain cells fall apart, and frequently found is a vacuolation of the CNS, known as spongiform change.
7. 'Inasmuch as our results are directly applicable to the biological situation, they suggest that the simple self-inductive mechanism which has been invoked by Monod and Jacob (1961) in connection with differentiation, and Griffith (1967) in relation to the disease Scrapie is only likely to give sufficient stability in the two states required if the induction is by co-operative mechanism with $m \geq 2$' (Griffith, 1968b: 215).

8. It should be noted that we use the phrases 'Prusiner and collaborators' and 'Prusiner's group' only to indicate that Prusiner appeared in the list of authors. Prusiner is not necessarily the first author of these articles, and the researchers involved may be from different institutions.
9. For details of this development, see Keyes (1999b).
10. On the different lines of research focusing on the disease manifestations and the molecular constitution of the agent, respectively, see Kim, Chapter 2 this volume.
11. Surely, we do not want to indicate that these are the only factors of importance in the development of the prion theory. Other important factors that form part of the generation and acceptance of the hypotheses of the prion theory are the molecularization of biology (see Kim, Chapter 2 this volume) and changed assumptions about the Central Dogma (see Keyes, 1999a,b).
12. Other authors have mentioned resemblances between the second mechanism for protein self-replication proposed by Griffith and the prion theory proposed by Prusiner. Some of these authors even point to Griffith's proposal as the original formulation of the concept of 'self-replication' of the prion protein (Keyes, 1999a; Laurent, 1997). Prusiner's own view of Griffith's work was expressed in, for example, a book published in 1999. Here, Prusiner acknowledged both the claims of Alper and of Pattison and recognized some aspects of Griffith's first and second proposals as truly predictive. However, Prusiner still reasoned:

    > Were Griffith's proposals of value in deciphering the prion problem? The answer is probably no, but his truly prescient speculations serve to enrich the history of the field. In contrast, Tikvah Alper's radiation inactivation data placed some important constraints on the physical features of the infectious pathogen of scrapie (Alper et al., 1966, 1967). Interestingly, many investigators ignored her findings and a few continue to do so although the results have been confirmed and greatly extended (Bellinger-Kawahara et al., 1987a,b, 1988). (Prusiner, 1999: 96)

13. See also Kim, Chapter 2 this volume, for a discussion of how the acceleration of molecular approaches to biological research questions made possible by the many new techniques developed in the 1980s and 1990s influenced research on the scrapie agent.
14. A few researchers have recognized Griffith's thermodynamical arguments. For example, Laurent has addressed the issue of the thermodynamic requirements allowing conformational change of the prion protein to occur, concluding that 'Thirty years later, our microscopic and mechanistic thermodynamical considerations agree with Griffith's original proposal' (Laurent, 1997: 6).
15. Comparing to Kim's (Chapter 2 this volume) distinction between generalists and experts, it is important to note that Prusiner had gained expertise in several different fields. Thus, his expertise should not be seen as narrow in scope as such.

References

Adams, D. H. and E. A. Caspary (1967) 'Nature of the Scrapie Virus', *British Medical Journal* 3: 173.

Adams, D. H. and E. J. Field (1968) 'The Infective Process in Scrapie', *Lancet* 2: 714–16.
Alper, T. et al. (1966) 'The Exceptionally Small Size of the Scrapie Agent', *Biochemical and Biophysical Research Communications* 22: 278–84.
Alper, T. et al. (1967) 'Does the Agent of Scrapie Replicate without Nucleic Acid?', *Nature* 214: 764–6.
Alper, T. et al. (1978) 'The Scrapie Agent: Evidence against its Dependence for Replication on Intrinsic Nucleic Acid', *Journal of General Virology* 41: 503–16.
Andersen, H. (1996) 'Categorization, Anomalies, and the Discovery of Nuclear Fission', *Studies in the History and Philosophy of Modern Physics* 27: 463–92.
Baltimore, D. (1970) 'Viral RNA-dependent DNA Polymerase', *Nature* 226: 1209–11.
Barker, P. et al. (2002) 'Kuhn on Concepts and Categorization', in T. Nickles (ed.) *Thomas Kuhn*. Cambridge: Cambridge University Press: 212–45.
Barry, R. A. et al. (1986) 'Scrapie and Cellular Prion Proteins Share Polypeptide Epitopes', *Journal of Infectious Diseases* 153: 848–54.
Basler, K. et al. (1986) 'Scrapie and Cellular Prp Isoforms Are Encoded by the Same Chromosomal Gene', *Cell* 46: 417–28.
Bastian, F. O. (1979) 'Spiroplasma-like Inclusions in Creutzfeldt-Jakob Disease', *Archives of Pathology and Laboratory Medicine* 103: 665–9.
Bellinger-Kawahara, C. et al. (1987a) 'Purified Scrapie Prions Resist Inactivation by Uv Irradiation', *Journal of Virology* 61: 159–66.
Bellinger-Kawahara, C. et al. (1987b) 'Purified Scrapie Prions Resist Inactivation by Procedures That Hydrolyze, Modify, or Shear Nucleic Acids', *Virology* 160: 271–4.
Bellinger-Kawahara, C. G. et al. (1988) 'Scrapie Prion Liposomes and Rods Exhibit Target Sizes of 55,000 Da', *Virology* 164: 537–41.
Bolton, D. C. and P. E. Bendheim (1988) 'A Modified Host Protein Model of Scrapie', *Ciba Foundation Symposium* 135: 164–81.
Bolton, D. C. et al. (1982) 'Identification of a Protein That Purifies with the Scrapie Prion', *Science* 218: 1309–11.
Cantor, G. N. (1975) 'The Edinburgh Phrenology Debate', *Ann. Sci.* 32: 195–218.
Chandler, R. L. (1959) 'Attempts to Demonstrate Antibodies in Scrapie Disease', *Veterinary Record* 71: 58–9.
Chandler, R. L. (1961) 'Encephalopathy in Mice Produced by Inoculation with Scrapie Brain Material', *Lancet* 1: 1378–9.
Chesebro, B. et al. (1985) 'Identification of Scrapie Prion Protein-Specific Messenger-RNA in Scrapie-Infected and Uninfected Brain', *Nature* 315: 331–3.
Cuillé, J. and P. L. Chelle (1936) 'La Maladie Dite Tremblante du Mouton. Est-elle Inoculable?', *Comptes Rendus Heb. des Séances de l'Académie des Sciences* 203: 1552–4.
Cuillé, J. and P. L. Chelle (1939) 'Transmission Expérimentale de la *Tremblante* à La Chèvre', *Comptes Rendus Heb. des Séances de l'Académie des Sciences* 208: 1058–60.
Darden, L. (1992) 'Strategies for Anomaly Resolution', in R. Giere (ed.) *Cognitive Models of Science*. Minnesota Studies in the Philosophy of Science Vol. XV, Minneapolis: University of Minnesota Press: 251–73.
Darden, L. (1998) 'Exemplars, Abstractions, and Anomalies: Representation and Theory Change in Mendelian and Molecular Genetics', in G. Wolters,

J. G. Lennox and P. McLaughlin (eds) *Concepts, Theories and Rationality in the Biological Sciences: the Second Pittsburgh–Konstanz Colloquium in the Philosophy of Science*. Konstanz/Pittsburgh: Universitätsverlag Konstanz/University of Pittsburgh Press: 137–58.

Diener, T. O. (1972) 'Is the Scrapie Agent a Viroid?', *Nature – New Biology* 235: 218–19.

Diener, T. O. (1973) 'Similarities between the Scrapie Agent and the Agent of the Potato Spindle Tuber Disease', *Annals of Clinical Research* 5: 268–78.

Eklund, C. M. et al. (1967) 'Pathogenesis of Scrapie Virus Infection in the Mouse', *Journal of Infectious Diseases* 117: 15–22.

Feyerabend, P. K. (1970) 'Consolations for the Specialist', in I. Lakatos and A. Musgrave (eds) *Criticism and the Growth of Knowledge*. Cambridge: Cambridge University Press: 197–230.

Field, E. J. (1966) 'Transmission Experiments with Multiple Sclerosis: an Interim Report', *British Medical Journal* 2: 564–5.

Gajdusek, D. C. and C. J. Gibbs (1968) 'Infections of the Nervous System', *Research Publications Association for Research in Nervous and Mental Disease* 44: 259.

Gajdusek, D. C. et al. (1966) 'Experimental Transmission of a Kuru-like Syndrome to Chimpanzees', *Nature* 209: 794–6.

Gibbons, R. A. and G. D. Hunter (1967) 'Nature of the Scrapie Agent', *Nature* 215: 1041–3.

Gibbs, C. J. and D. C. Gajdusek (1969) 'Infection as the Etiology of Spongiform Encephalopathy (Creutzfeldt-Jakob Disease)', *Science* 165: 1023–5.

Gibbs, C. J. et al. (1968) 'Creutzfeldt-Jakob Disease (Spongiform Encephalopathy): Transmission to the Chimpanzee', *Science* 161: 388–9.

Gordon, W. S. (1946) 'Advances in Veterinary Research', *Veterinary Record* 58: 516–20.

Greig, J. Russell (1950) 'Scrapie in Sheep', *Journal of Comparative Pathology* 60: 263–6.

Griffith, J. S. (1967) 'Self-replication and Scrapie', *Nature* 215: 1043–4.

Griffith, J. S. (1968a) 'Mathematics of Cellular Control Processes. I. Negative Feedback to One Gene', *Journal of Theoretical Biology* 20: 202–8.

Griffith, J. S. (1968b) 'Mathematics of Cellular Control Processes. II. Positive Feedback to One Gene', *Journal of Theoretical Biology* 20: 209–16.

Hadlow, W. J. (1959) 'Scrapie and Kuru', *Lancet* 2: 289–90.

Haig, D. A. et al. (1969) 'Further Studies on the Inactivation of the Scrapie Agent by Ultraviolet Light', *Journal of General Virology* 5: 455–7.

Hay, B. et al. (1987) 'Evidence for a Secretory Form of the Cellular Prion Protein', *Biochemistry* 26: 8110–15.

Hsiao, K. et al. (1989) 'Linkage of a Prion Protein Missense Variant to Gerstmann–Straussler Syndrome', *Nature* 338: 342–5.

Hull, D. et al. (1978) 'Planck's Principle. Do Younger Scientists Accept New Scientific Ideas with Greater Alacrity than Older Scientists?', *Science* 202: 717–23.

Hunter, G. D. (1969) 'The Size and Intracellular Location of the Scrapie Agent', *Biochemical Journal* 114: 22P–23P.

Hunter, G. D. (1992) 'The Search for the Scrapie Agent: 1961–1981', in S. B. Prusiner et al. (eds) *Prion Diseases of Humans and Animals*. New York: Ellis Horwood: 23–9.

Hunter, G. D. and G. C. Millson (1964) 'Studies on the Heat Stability and Chromatographic Behaviour of the Scrapie Agent', *Journal of General Microbiology* 37: 251–8.
Keyes, M. E. (1999a) 'The Prion Challenge to the "Central Dogma" of Molecular Biology, 1965–1991. Part I: Prelude to Prions', *Studies in the History and Philosophy of Biological and Biomedical Sciences* 30: 1–20.
Keyes, M. E. (1999b) 'The Prion Challenge to the "Central Dogma" of Molecular Biology, 1965–1991. Part II: The Problem with Prions', *Studies in the History and Philosophy of Biological and Biomedical Sciences* 30: 181–218.
Klatzo, I. et al. (1959) 'Pathology of Kuru', *Laboratory Investigation* 8: 799–847.
Kuhn, T. S. (1962) *The Structure of Scientific Revolutions*, 2nd edn 1970. Chicago: University of Chicago Press.
Lakatos, I. (1970) 'Falsification and the Methodology of Research Programmes', in I. Lakatos and A. Musgrave (eds) *Criticism and the Growth of Knowledge*. Cambridge: Cambridge University Press: 91–196.
Laudan, L. (1977) *Progress and Its Problems*. Berkeley: University of California Press.
Laurent, M. (1997) 'Autocatalytic Processes in Cooperative Mechanisms of Prion Diseases', *FEBS Letters* 407: 1–6.
M'Fadyean, J. (1918) 'Scrapie', *Journal of Comparative Pathology and Therapeutics* 31: 102–31.
M'Gowan, J. P. (1914) *Investigation into the Disease of Sheep called 'Scrapie'*. Edinburgh: William Blackwood.
Mckinley, M. P. et al. (1983) 'A Protease-Resistant Protein Is a Structural Component of the Scrapie Prion', *Cell* 35: 57–62.
McMullin, E. (1992) 'Rationality and Paradigm Change in Science', in P. Horwich (ed.) *World Changes*. Cambridge, Mass.: MIT Press: 55–80.
Masiarz, F. R. et al. (1980) 'Purification of the Scrapie Agent–Strategy and Recent Progress', in A. Boese (ed.) *Search for the Cause of Multiple Sclerosis and other Chronic Diseases of the Central Nervous System*. Weinheim: Verlag Chemie: 321–32.
Messeri, P. (1988) 'Age Differences in the Reception of New Scientific Theories: the Case of Plate Tectonics', *Social Studies of Science* 18: 91–112.
Narang, H. K. (1974) 'Ruthenium Red and Lanthanum Nitrate a Possible Tracer and Negative Stain for Scrapie "Particles"?', *Acta Neuropathologica* 29: 37–43.
Oesch, B. et al. (1985) 'A Cellular Gene Encodes Scrapie Prp 27–30 Protein', *Cell* 40: 735–46.
Oesch, B. et al. (1988) 'Search for a Scrapie-specific Nucleic Acid: a Progress Report', *Ciba Foundation Symposium* 135: 209–23.
Parry, H. B. (1962) 'Scrapie: a Transmissible and Hereditary Disease of Sheep', *Heredity* 17: 75–105.
Pattison, I. H. (1965) 'Resistance of the Scrapie Agent to Formalin', *Journal of Comparative Pathology* 75: 159–64.
Pattison, I. H. (1966a) 'Experiments with Scrapie with Special Reference to the Nature of the Agent and the Pathology of the Disease', in D. C. Gajdusek et al. (eds) *Slow, Latent, and Temperate Virus Infections*. NINDB monograph No. 2. Public health service publication: 249–57.

Pattison, I. H. (1966b) 'The Relative Susceptibility of Sheep, Goats and Mice to Two Types of the Goat Scrapie Agent', *Research in Veterinary Science* 7: 207–12.

Pattison, I. H. (1970) 'Recent Work on Scrapie', *Pathologie Biologie* 18: 673–8.

Pattison, I. H. (1988) 'Fifty Years with Scrapie: a Personal Reminiscence', *Veterinary Record* 123: 661–6.

Pattison, I. H. (1992) 'A Sideways Look at the Scrapie Saga: 1732–1991', in S. B. Prusiner et al. (eds) *Prion Diseases of Humans and Animals*. New York: Ellis Horwood: 15–22.

Pattison, I. H. and K. M. Jones. (1967) 'The Possible Nature of the Transmissible Agent of Scrapie', *Veterinary Record* 80: 2–9.

Pattison, I. H. and G. C. Millson (1960) 'Further Observations on the Experimental Production of Scrapie in Goats and Sheep', *Journal of Comparative Pathology and Therapeutics* 70: 182–93.

Pattison, I. H. and Millson, G. C. (1961a) 'Further Experimental Observations on Scrapie', *Journal of Comparative Pathology and Therapeutics* 71: 350–9.

Pattison, I. H. and G. C. Millson (1961b) 'Scrapie Produced Experimentally in Goats with Special Reference to the Clinical Syndrome', *Journal of Comparative Pathology and Therapeutics* 71: 101–8.

Penrose, L. S. and R. Penrose (1957) 'A Self-reproducing Analogue', *Nature* 179: 1183.

Prusiner, S. B. (1982) 'Novel Proteinaceous Infectious Particles Cause Scrapie', *Science* 216: 136–44.

Prusiner, S. B. (1984) 'Prions', *Scientific American* 251: 48–57.

Prusiner, S. B. (1991) 'Molecular-Biology of Prion Diseases', *Science* 252: 1515–22.

Prusiner, S. B. (1995) 'The Prion Diseases', *Scientific American* 272: 30–7.

Prusiner, S. B. (1999) 'Development of the Prion Concept', in S. B. Prusiner (ed.) *Prion Biology and Diseases*. New York: Cold Spring Harbor Laboratory Press: 67–112.

Prusiner, S. B. et al. (1981) 'Scrapie Agent Contains a Hydrophobic Protein', *Proceedings of the National Academy of Sciences of the United States of America – Biological Sciences* 78: 6675–9.

Prusiner, S. B. et al. (1982) 'Further Purification and Characterization of Scrapie Prions', *Biochemistry* 21: 6942–50.

Prusiner, S. B. et al. (1990) 'Transgenetic Studies Implicate Interactions between Homologous PRP Isoforms in Scrapie Prion Replication', *Cell* 63: 673–86.

Rappa, M. and K. Debackere (1993) 'Youth and Scientific Innovation: the Role of Young Scientists in the Development of a New Field', *Minerva* 31: 1–20.

Rohwer, R. G. and D. C. Gajdusek (1980) 'Scrapie – Virus or Viroid. The Case for a Virus', in A. Boese (ed.) *Search for the Cause of Multiple Sclerosis and other Chronic Diseases of the Central Nervous System*. Weinheim: Verlag Chemie: 333–55.

Scott, M. et al. (1989) 'Transgenic Mice Expressing Hamster Prion Protein Produce Species-Specific Scrapie Infectivity and Amyloid Plaques', *Cell* 59: 847–57.

Sigurdsson, B. (1954) 'Rida, a Chronic Encephalitis of Sheep', *British Veterinary Journal* 110: 341–54.

Stamp, J. T. (1962) 'Scrapie: a Transmissible Disease of Sheep', *Veterinary Record* 74: 357–62.

Stamp, J. T. (1967) 'Scrapie and its Wider Implications', *British Medical Bulletin* 23: 133–7.

Stockman, S. (1926) 'Contribution to the Study of the Disease Known as Scrapie', *Journal of Comparative Pathology and Therapeutics* 39: 42–71.

Temin, H. M. and S. Mizutani (1970) 'RNA-dependent DNA-polymerase in Virions of Rous Sarcoma Virus', *Nature* 226: 1211–13.

Toulmin, S. (1972) *Human Understanding*, Vol. 1. Princeton, NJ: Princeton University Press.

Yost, C. S. et al. (1990) 'Non-hydrophobic Extracytoplasmic Determinant of Stop Transfer in the Prion Protein', *Nature* 343: 669–72.

2
Styles of Scientific Practice and the Prion Controversy

Ki-Heung Kim

Introduction

In 1982, Stanley Prusiner proposed the prion hypothesis to account for the unusual characteristics of the agent of transmissible spongiform encephalopathies (TSEs). However, his was not the only alternative hypothesis that sought to explain the strange behaviour of this infectious agent. At about the same time Alan Dickinson, later director of the Neuropathogenesis Unit (NPU) in Edinburgh, suggested that the agent might be a virino, that is, a piece of nucleic acid coated with host protein. These opposing views led to a controversy between two groups of TSE researchers.

In order to analyse this controversy, I will use a theoretical framework 'styles of scientific practice', which is extended from Jonathan Harwood's work on style. In the course of the controversy, divergent ideas derived not only from actual experimental results, but also from different styles of research programmes determined by different styles of practice. I will describe how the controversy proceeded during the 1980s and 1990s. I will analyse disputes over several significant issues such as the prion gene and infectivity of the prion protein, transgenic experiments, and the biological diversity of the agent. I will show that the way in which these issues were addressed by the two groups was fundamentally related to their respective style of scientific practice.

This chapter will also deal with why both groups of scientists adhered passionately to their own blend of scientific programme, and why they could not reach any agreement on the nature of the agent. I will show that the controversy was brought about by the confrontation of prion sceptics' 'generalist biological' style and Prusiner's 'specialist biochemical' style. The former focuses upon the whole mechanism of disease and the host, whereas the latter centres on particular molecules. The distinction between generalist and specialist styles is invoked by practitioners themselves to justify their respective way of studying TSEs. This difference of style has produced totally distinctive experimental systems and results. Even though each set of experimental techniques has produced valuable knowledge, the two camps still disagree on the significance of their adversaries' achievements. From the viewpoint of medical history, this controversy cannot be explained by factors such as the quality of the empirical evidence produced. Rather, I will show that in science experimental systems cannot be dissociated from styles of practice. Furthermore, I will show that one of these styles of research programme is better adapted to contemporary developments in biomedicine. This is vital to understand why Prusiner's molecular biological research programme eventually gained more credibility within the scientific community, even though the prion controversy is as yet unsettled.

Styles of scientific practice

Traditionally, the concept of style has been a tool for classifying cultural patterns in the history of art. However, some sociologists attempted to apply the concept to various sociocultural phenomena. For instance, the sociologist Karl Mannheim adopted this concept to identify a variety of social groups' articulated thought (Mannheim, 1953). Mannheim addressed the question of why a specific style of thought with specific features is associated with a particular social group (for example, class, clan, nation and religious group) in a certain context. However, he exempted scientific knowledge from his sociological analysis.

More recently, a significant piece of work was produced by Jonathan Harwood. Harwood analyses how differently patterned cultures emerge and are maintained. His work embraces historicity of styles and coexistence of different styles of thought (Harwood, 1993). Harwood exemplifies his theory by focusing on the development of genetics in Germany in the early twentieth century. He analyses the development of genetic research, and compares it to the social, educational and institutional background of different research communities. He remarks that national differences of scientific traditions were clearly maintained in scientists' practice. Furthermore, he demonstrates that different cognitive patterns associated with scientists' social background can be identified within a particular national context (Harwood, 1987).

In his work on different types of genetics in Germany and the USA, Harwood emphasizes the concept of styles and claims that this concept is not intended to contrast the ontological or epistemological underpinnings of German and American genetics, but to reveal the range of questions that geneticists in the two countries took to be central to their discipline (Harwood, 1987: 391). Thus, he shows how different styles of scientific thought were generated in different social circumstances. For instance, various socio-economic structures of German society between the world wars reinforced German geneticists' theory-oriented approach and pushed them towards developmental biology. In contrast, with the rapid division of academic disciplines in the USA, American geneticists tended to address more narrowly defined problems and to develop an experiment-based specialist programme. Harwood claims that these stylistic differences stemmed not only from different university and funding structures, but also from different social and economic conditions after the First World War.

In this chapter, I extend Harwood's work on styles of thought to the domain of scientific practice. Most analyses on style have so far focused on reasoning and thought while practice is neglected. And yet, the domain where scientific theory is constructed and every actor's interests are created and maintained is practice. This shift of focus from scientific thought to scientific practice is also found in Joan Fujimura and Danny Chou's work (Fujimura and Chou, 1994, 1995). They argue that the debate on the origin of AIDS can be elucidated in terms of patterns of practice, and they conceptualize different styles of scientific practice. Fujimura and Chou explain that 'styles of practice are historically located and collectively produced work processes, methods, and rules for verifying theory. [...] Style of practice implies that practices of theory construction, adjudication, and maintenance are situated actions' (Fujimura and Chou, 1994: 1020–1). Based on Harwood and Fujimura and Chou's inspirational arguments, we will see that in the prion controversy different styles of practice, and in particular different styles of research programme, pervaded scientists' conduct of experiments, interpretation of data, use of techniques, construction of models, and the organizational structure of their research programmes. In the next section, I will describe the opening stages of the controversy.

Prions: proteinaceous infectious particles

In 1982, a neurologist of the University of California in San Francisco, Stanley Prusiner, published a review paper in *Science*. After summarizing his predecessors' experimental data and his own biophysical work, Prusiner proposed a radical speculation on the nature of the scrapie agent (Prusiner, 1982a). He concentrated his attention on the biophysical characteristics of scrapie and concluded that the attempts to find the scrapie-specific genome were unlikely to be successful. His own experiments with various chemico-enzymatic treatments of infected material such as D-Nase, R-Nase and protease-K, had failed to reveal it. However, from his experimental results, the only thing he could argue was that proteins were the main component of the infectious agent. If DNA was the main component of the agent, D-Nase would digest its DNA structure. Yet, his experiments with D-Nase had failed to inactivate the agent. On the other hand, when the agent was treated with protease-K, which *could* digest the amino-acid chain of proteins, the agent was inactivated.

From these experiments, Prusiner outlined two possible models of the structure of the agent. One was a small putative nucleic acid wrapped by a tightly packed protein coat. Another possible explanation was a protein devoid of nucleic acid. Although he presented two possible structures of the agent, he inclined towards the second option. Importantly, Prusiner used the opportunity of his review article to suggest a different name for the scrapie agent, which ignored any supposed viral characteristics and pointed to particular biochemical properties. Prusiner proposed calling the agent a 'prion':

> Because the dominant characteristics of the scrapie agent resemble those of a protein, an acronym is introduced to emphasise this feature. In place of such terms as 'unconventional virus' or 'unusual slow virus-like agent', the term 'prion' (pronounced *pree-on*) is suggested. Prions are small *pro*teinaceous *in*fectious particles, which are resistant to inactivation by most procedures that modify nucleic acids. The term 'prion' underscores the requirement of a protein for infection. (Prusiner, 1982a: 141)

It should be stressed that this new name set Prusiner apart from other researchers in two respects. First, in refusing to use such terms as 'unconventional slow virus' or 'virino', Prusiner was distancing himself from any supposition that the scrapie agent was similar to viruses. And secondly, in defining prions as 'small proteinaceous particles', he clearly underscored the essential role of proteins in the infectivity of the scrapie agent.

In his review article, Prusiner did not say that prions consisted solely of protein, that is, he did not rule out the possibility that nucleic acids were somehow involved in prion replication. However, in calling the agent a prion, and thereby emphasizing the functional importance of proteins in its action, Prusiner was playing up the importance of his own research since he had established that functional importance.

An interesting feature of Prusiner's article is that instead of putting forward a complete hypothesis on how the agent might replicate in the absence of a nucleic acid genome, he only coined a new name. At the time many scientists involved in scrapie research were attempting to explain the perplexing characteristics of the agent with their own hypothetical ideas. Prusiner was more cautious and more astute since acceptance of the term 'prion' did not depend upon the truth of any complete theoretical speculations. This term could serve to denote the scrapie agent irrespective of its chemical nature. Richard Carp, a leading prion sceptic at the Institute of Basic Research (IBR) in New York, says:

> We felt that certainly initially Prusiner attempted to maintain the term in such a way that even if nucleic acid was there, the term would still be applicable. In other words, the initial definitions did not rule out the possibility that the nucleic acid was there. And only subsequently did he change that definition such that it was protein only. But, it seemed initially, in my mind, that he wanted to have it all ways; that if a nucleic acid turned up in the next day, he could say, 'well, it is a prion', and if it didn't, then he could say, 'it is a prion'. That was my general feeling. (Carp, 2000)

Nevertheless, many researchers were critical of Prusiner's strategy. Firstly, the timeliness of his new classificatory term, prion, was questioned. Carleton Gajdusek, 1976 Nobel laureate for his work on kuru, pointed out that Prusiner's new naming of TSE agents was premature because no one was sure of their biochemical nature (Rhodes, 1997a: 161–3). Secondly, many TSE researchers construed prions as 'self-replicating proteins'. This meant, as Prusiner himself noted in his review paper, that there was a danger of contradicting the conventional wisdom of molecular biology known as the 'Central Dogma' (Crick, 1970). For this reason, in the early 1980s Prusiner's idea became a biological heresy among scientists.

Prusiner denied that he was arguing that protein was the only constituent of the infectious agent; however, even the first public appearance of the prion in the *San Francisco Chronicles* described it as such. The newspaper claimed that the prion cannot contain enough genetic material to reproduce itself (Perlman, 1982). Fortunately for Prusiner, an abnormal form of a protein was found just a month later. One of his postdoctoral researchers, David Bolton, found in scrapie-infected material a small protein component that was not digested by protease-K. Normally, every protein should be denatured when treated with this highly active non-specific protease (Bolton, 2000), and it was very commonly used to digest proteins at that time. It was therefore extraordinary that a scrapie-specific protein was not digested by protease-K (Bolton et al., 1982). Bolton and colleagues named this protein 'protease resistant protein (PrP)' or PrP 27-30, due to its molecular weight. Since then Prusiner's group have thought that this PrP form was the purified form of the infectious agent.

At the end of the 1970s, leading scrapie researchers in Edinburgh had suggested the virino hypothesis to account for the nature of the scrapie agent (Dickinson and Outram, 1979). Since the early 1960s, Edinburgh scientists had accumulated a vast amount of biological data. They had isolated several strains of the scrapie agent characterized by different pathological patterns and incubation periods. From these data, Alan Dickinson and his colleagues in Edinburgh concluded that scrapie behaves as an independent pathogen with its own coded information. Not surprisingly, soon after the publication of Prusiner's 1982 review paper, vehement criticisms were orchestrated by two of these scrapie researchers now based at the Neuropathogenesis Unit (NPU) in Edinburgh.[1]

In a paper published in *Nature*, Richard Kimberlin put forward: 'scrapie is naturally infectious; the disease can be transmitted experimentally to many different species including primates and rodents; several genetically stable strains of agent have been isolated, strain selection can occur on serial passage of mixtures and some strains show properties which strongly suggest mutations in the scrapie genome' (Kimberlin, 1982a: 393). Instead of using Prusiner's term, prion, Kimberlin thought that the neologism 'virino' suggested by Dickinson and Outram, would be preferable for describing the possible character of protein-wrapped scrapie-specific genome.

A few weeks later, in an anonymous editorial published in the *Lancet*,[2] Dickinson stated that Prusiner's novel idea was premature as shown by the fact that many people in the 1930s wrongly assumed that conventional viruses consisted essentially of protein. Furthermore, Prusiner's idea of the possible absence of a scrapie-specific genome could not account plausibly for the 'various ramifications of the occurrence of different strains of scrapie' (Anonymous [Dickinson], 1982: 1222).

Shortly after the *Lancet* editorial was published, Prusiner recognized the style of Dickinson's writing, and Dickinson 'got a rocket', as he put it, from Prusiner (Dickinson, 1999b). In a letter published in the

Lancet (Prusiner, 1982b), Prusiner expressed his scepticism about the Edinburgh group's 20-year achievements, and questioned the value of their work on strain variation of the scrapie agent. He argued: 'to suggest that isolation of a few strains of the scrapie agent in laboratory rodents for pathogenesis studies is an important achievement is questionable. These strains may describe a few biological characteristics of the scrapie agent, but they do not define or constrain the possible molecular structures of this unusual infectious particle' (Prusiner, 1982b: 494). This volatile confrontation was a sort of declaration of war against the Edinburgh group. In criticizing the existence of strain variations directly, Prusiner was crossing the Rubicon with regard to attacking mainstream studies. His challenge was the first salvo in a long warfare between the prion group and prion sceptics.

The prion controversy

Soon after the exchange in *Nature* and *Lancet*, the battlelines were clearly drawn between Prusiner's group and prion sceptics. In the early 1980s, at least eight leading groups of scientists were studying TSE agents.[3] The majority of TSE researchers belonged to the sceptics' faction; however, it should be stressed that prion sceptics were not a homogeneous group. Sometimes they shared the same ideas on the nature of the agent, but sometimes they took different positions. Thus, the term 'prion sceptics' designates those who fundamentally disagreed with Prusiner's notion of a protein-only infectious agent. The leading group of sceptics was at the NPU. Their experimental demonstration of strain variation was the most powerful scientific evidence to persuade others that the agent could be classified as virus-like.

Almost everyone involved in TSE research became Prusiner's adversary. His new idea was considered 'biological heresy' and some dismissed his alleged prion theory as simply 'a fairy tale'. According to science writer Jennifer Cooke, 'his many detractors at the time labelled Prusiner, and his heresy, as "the P words" ... he had created a new scientific word to fit a scientific entity that was still unknown. And for that he attracted a lot of publicity – a third "P" word which resulted in grant money' (Cooke, 1998: 106). Even Prusiner's former collaborator, Richard Race, did not believe in his concept and its theoretical implications. He says: 'it was heretical, this is nuts. It is crazy. But over the years my attitude is maybe it is, but we need better evidence' (Race, 2000).

Through the 1980s and the 1990s, the two factions of scientists conducted various complicated experiments to prove their respective theory. Whenever experimental results were published, the competitors produced counter-arguments. In the prion controversy, there were several important issues. In the following sections, I will describe how contenders conducted experiments and raised their arguments during the controversy. I will start with the discovery of the prion protein gene in 1985.

Prion protein gene (1985)

Having purified the protein PrP as mentioned above (Bolton et al., 1982), Prusiner was in a position to further investigate its role in scrapie infection. Prusiner was particularly interested in testing his hypothesis that PrP was itself the agent, and that it propagated by catalysing its own manufacture in the cell. This implied that no evidence could be found that PrP was being manufactured by translation from either the host genome or a viral genome. If he could rule out genetic translation, this would place him in a much stronger position to suggest that the scrapie agent is a self-replicating protein. Hence, Prusiner devised a series of experiments to test whether or not there was any evidence of PrP being manufactured by normal genetic means in infected hamster brains. To carry out this work, Prusiner looked beyond the confines of the scrapie research community and drew on the expertise of Leroy Hood of Caltech and Charles Weissmann of the University of Zurich. These two scientists were known among medical students as 'gods of molecular biology' (Taubes, 1986: 50). Hood was a pioneer in cloning techniques for sequencing DNA. He was also one of the first advocates of, and a key player in, the Human Genome Project. Weissmann was an expert in gene cloning and gene splicing (Lane, 1997). Weissmann accepted Prusiner's suggestion of collaboration because he had attended Prusiner's talk on scrapie in Perth, Australia, in 1982 (Brown, 1999: R625) and since then was fascinated by scrapie. Crucially, Prusiner devised his series of experiments with Hood and Weissmann as a continued programme of research, though the experiments were conducted separately. Hood was involved in isolating short amino-acid chains from prion proteins, whereas Weissmann dealt with cloning these amino-acid chains and isolating nucleic acid from them.

Molecular biological orthodoxy states that DNA sequences in the genes provide a template for the transcription of so-called complementary DNA (cDNA). This in turn provides a template for transcribing messenger RNA (mRNA). And messenger RNA in turn provides a template for reading off the sequence of amino acids that make up a protein. Prusiner's aim was to verify whether it was possible to identify any polynucleotides in scrapie-infected brain material that might correspond to these stages in the manufacture of PrP.

In its collaborative work with Hood, Prusiner's team successfully produced a number of fairly short amino-acid chains, called oligopeptides, from PrP in 1984. These were particular sequences of amino acids which, taken together, would almost certainly be unique to PrP. Knowing these amino-acid sequences, the researchers were then able to work backwards to specify what nucleotide sequences in mRNA would code for them (Prusiner et al., 1984: 132). However, the problem was that there is redundancy in the coding, that is, most amino acids can be coded by more than one nucleotide sequence. Consequently, working backwards from oligopeptides to nucleotide sequences led to the specification of a rather large number of mRNA sequences, each of which could code for the respective oligopeptide chains of PrP. Prusiner thus undertook the rather laborious task of chemically manufacturing the various candidate oligonucleotide sequences. Prusiner needed to find out if any of these candidate mRNA sequences was identical to mRNA that actually occurs in cells where scrapie is replicating and hence where PrP is being manufactured. If he *could* find such mRNA *in vivo*, then it would show that PrP is synthesized from genetic information, and this would refute his self-replicating protein hypothesis.

As a result of this experimental project, the collaborative team of Prusiner and Weissmann found that some of his PrP-specific mRNA did indeed bind to the *E. coli* genome in some of his cultures, that is, these mRNA sequences were indeed complementary to cDNA derived from scrapie-infected hamsters. Contrary to what he expected, Prusiner had demonstrated that PrP was manufactured in scrapie-infected hamster cells by a process of translation of information coded in a nucleic acid genome. It should be noted, though, that at this stage Prusiner still did not know exactly where this cDNA originated in scrapie-infected hamster cells. It could be transcribed from the hamster's own genome, or from an exogenous genome, for instance one belonging to a putative scrapie virus. Prusiner and colleagues eventually showed that this gene is present in non-infected as well as in infected hamster cells. In other words, it is part of the normal hamster genome (Oesch et al., 1985).

Of course, all this constituted strong evidence against Prusiner's original theory that the scrapie agent is a self-replicating protein, since he had shown that PrP, the protein he considered most likely to play this role, was a product of the hamster's own genes. It took Prusiner and colleagues six months to make sense of this 'self-inflicted apparent disproof of his theory' (Taubes, 1986: 50). Bolton, a Prusiner researcher at that time, says: 'the impression I get is that they were quite perplexed about this gene showing up. If you read the articles, Weissmann seems uncomfortable with how you mesh this being a normal gene with it being a prion' (Bolton, 2000).

In conclusion, Prusiner's own experimental results effectively disproved his initial hypothesis that the scrapie agent is simply a self-replicating protein associated with, and perhaps even identical to, PrP. He nonetheless managed to salvage something of his protein hypothesis by raising further interesting questions about the role of PrP in scrapie infection. The prion protein gene was shown to produce both the normal cellular isoform of PrP (PrP^C) and the scrapie pathological isoform (PrP^{Sc}) of the protein, and after speculating for some time Prusiner and Weissmann suggested that the protease-resistant, disease-associated isoform PrP^{Sc} was responsible for scrapie-like diseases. Though Prusiner may not yet have been in a position to formulate this view explicitly, it appears that he was already entertaining a suspicion that infection might proceed through the conversion of PrP^C into PrP^{Sc} (Basler et al., 1986).

Is PrP^{Sc} infectious? Counter-evidence (1986–90)

In spite of Prusiner's impressive technical accomplishments in identifying and isolating PrP and the gene that codes for it, many scientists remained sceptical about the theoretical conclusions he drew from his work, and especially about his claim that PrP was a key factor in scrapie infectivity. Furthermore, prion sceptics continued to conduct research that appeared to support their scepticism.

At the same time as Prusiner was working on the PrP gene, a group of scrapie researchers at the Rocky Mountain Laboratory (RML) led by Bruce Chesebro, launched a similar project to scrutinize the troublesome protein, including the quest for the mouse gene that codes for it (Chesebro et al., 1985). Their experimental design was nearly the same as that of Prusiner's team. With the same small amino-acid sequences from scrapie protein that were isolated by Prusiner and Hood's team in 1984, Chesebro's team synthesized a mixture of oligonucleotides for use as a hybridization probe to analyse mRNA populations derived from infected and uninfected animals. As a result of this investigation, like Prusiner they concluded that the gene for the normal and pathological proteins was identical, and that there was no evidence for any unique mRNA associated with scrapie infectivity. And they drew possible implications from this finding, though different from Prusiner's. Firstly, it could be seen as evidence that scrapie infection was not caused by the host gene. Secondly, if expression of the prion protein is associated with scrapie, Chesebro and his team speculated that it could be due to a post-transcriptional or mutational modification (Chesebro et al., 1985: 332).

More interestingly, Chesebro's team failed to find any PrP-specific mRNA in mouse spleen. PrP-specific mRNA appeared only in scrapie-affected brain, neither in the spleen nor in the liver. This was puzzling since many pathogenetic studies of scrapie had shown that the agent replicates first in the spleen (Kimberlin, 1979; Outram, 1976). If PrP was really the infective agent, why was PrP-related mRNA not found in the spleen? Chesebro concluded that PrP 27-30, which was the candidate infectious agent for Prusiner's group, might not be specific to the infectious scrapie agent (Chesebro et al., 1985: 332). This experimental result cast doubt on Prusiner's 1985 observations.

Shortly afterwards, Laura Manuelidis, a worldwide CJD expert at the Yale Medical School, came up with an interesting experimental result showing that the prion protein may not be linked to scrapie infectivity (Manuelidis et al., 1987). Manuelidis' team set out to reassess the effect of protease-K treatment on scrapie infectivity. Other researchers had already shown that protease-K significantly reduced scrapie infectivity (Millson et al., 1976; Lax et al., 1983), but Prusiner's claims rested on the assumption that the infectious agent resists protease-K treatment. Manuelidis now showed that protease treatment of CJD brain material does indeed produce a protease-K resistant form of PrP as Prusiner's team observed. But at the same time, Manuelidis also observed that this treatment reduces infectivity by more than 90 per cent (Manuelidis et al., 1985).

Furthermore, Manuelidis found that in CJD preparations, the major protein equivalent to PrP 27–30 in Prusiner's scrapie experiments could be separated from infectivity under mild non-denaturing conditions, whereas Prusiner's group suggested that the infectivity was inseparable from PrP. Also, Manuelidis and colleagues later reported that when they attempted to separate different molecules from infected brain samples, the most infectious part was not PrP but a fraction containing other proteins and nucleic acids (Sklaviadis et al., 1989). These studies suggested that PrP in itself was unlikely to be the replicating component of the infectious agent and Manuelidis claimed that the agent was virus-like and had its own informational molecule. Many viruses, she argued, are hardy and even resist treatment with enzymes that digest genetic material. These viruses like the poliovirus are packed inside a protein shell that protects them. 'Think of all the viruses out there that have to get through the gastrointestinal tract,' she said, 'they have to deal with all sorts of lousy environments' (Kolata, 1994: C1).

In 1989, more good news for the sceptics came from Edinburgh. At the time prion sceptics were pointing out that Prusiner's suggestion that the scrapie agent is an infectious protein seemed unable to account for the significant and problematic fact of strain variation in scrapie. This objection was reinforced when Kimberlin and colleagues published new work on strain variation (Kimberlin et al., 1989). They studied the transmission of different strains of scrapie from mice to hamsters and then back to mice. In this experiment, each strain of the agent maintained its distinctive pathogenic identity when the agent was transferred between different species like mice and hamsters. Kimberlin showed that in spite of the species barrier, the scrapie agent maintains its genomic character when it jumps to a new species. From this experiment, Kimberlin speculated that 'it is likely that the scrapie genome is a very small "regulatory" nucleic acid which may not code for protein (hence the need for a "protective" host-coded protein such as PrP). A major criterion for recognising candidate genomes is that there should be sequence differences according to the strain of agent' (Kimberlin et al., 1989: 2018). In other words, while the scrapie agent may rely on the host genome to manufacture its protein constituents, the strain is not determined by the gene of the infected animal but by the infective agent (Dealler, 1996: 52). This was at odds with Prusiner's hypothesis that the scrapie agent is simply a protein coded by the host genome, and further reinforced the view that the agent must have a genome of its own. Indeed, Kimberlin actually included the phrase 'genomic identity' in the title of his 1989 paper.

As a result of these experimental findings, prion sceptics remained unconvinced and pointed both to Prusiner's failure to link PrP unequivocally to infectivity, and to his failure to account for strain variation and strain conservation between different species. Meanwhile, as more and more energy and resources were invested in scrapie research, the stakes grew higher and between 1986 and 1989, the scrapie research community was embroiled in their most bitter clashes in the history of the prion controversy. Each camp presented experimental achievements to refute their opponents. However, the dispute was moving beyond a rational debate. Throughout these experimental exchanges, the relationship between the two sides was becoming increasingly acrimonious. For instance, at a CIBA Foundation meeting in 1987, there was a major confrontation and the controversy between the prion group and prion sceptics dramatically intensified. It was so intense that the meeting organizers stopped recording the proceedings. George Carlson says: 'I was amazed, I never saw anything like that. The personalities were amazing. You had people yelling at each other at meetings…I mean it is mind-boggling, absolutely mind-boggling. The animosity between groups, it was very controversial' (Carlson, 2000).

Transgenic experiments and strain variation

Strain variation is the most controversial issue in the history of TSE research. For Prusiner's framework, it is the weakest link. Since 1965, Dickinson and co-workers in Edinburgh have established this concept on the basis of observations of various incubation periods and pathological changes of scrapie-infected mice. For instance, two distinct lines of symptoms were maintained which produced strikingly different clinical signs in goats from the same herd, either a 'drowsy' or 'scratching' syndrome (Pattison and Millson, 1961). Dickinson and Hugh Fraser have serially passaged scrapie from a wide range of natural and experimental sources in inbred mice. With these methods, Edinburgh researchers were able to isolate as many as 20 strains of scrapie (Dickinson et al., 1968; Dickinson and Meikle, 1969; Dickinson, 1976; Bruce and Fraser, 1991). Since then, they have developed quantitative methods for estimating different clinical characteristics. Their research has provided vital data highlighting the viral characteristics of scrapie.

During the early days of the controversy, Prusiner really discounted the reality of strains. David Bolton states that 'he really thought that it was not significant' (Bolton, 2000). Richard Carp (IBR) comments critically: 'I think in the mid-1980s, the San Francisco group didn't believe there were any strains. There were just two strains, a mouse strain and hamster strain. It was only subsequently that they believed in strain variation in hamsters and mice. Initially they didn't believe that' (Carp, 2000). Prusiner's denial of strain variation ignited the heated debate with Edinburgh researchers. In a review paper he repeatedly challenged the reality of strains and claimed that 'the purity of these isolates is unknown since plaque purification methods are unavailable…if a large number of strains of the scrapie agent exist as suggested, the most plausible model for the prion would seem to require a nucleic acid. There is no evidence to date for such a structure' (Prusiner, 1984: 22).

Indeed, strain variation had an explosive potential for refuting Prusiner's protein-only idea. If strain variation could not be explained by the prion framework, then many would incline to believe in the virino theory. Moreover, when Prusiner and his collaborators in Zurich found that the prion protein was coded by a normal prion gene, his early hypothesis on the self-replicating protein was on the brink of collapse. The following statement by one of Prusiner's colleagues in

San Francisco, David Westaway, may be a good example of how prion people felt frustrated with the controversy on strain variation: 'I am not really happy that these strain types are actually as perfect as they're putting across anyway. All I can say for the moment is that there must indeed be some methods by which the protein retains its strain. But just because I don't understand it doesn't mean that it's not true' (Dealler, 1996: 55).

The strain variation controversy was constantly refuelled by novel experimental results from both camps. In 1989, Prusiner's team achieved a new breakthrough by using the transgenic technique, which was then an advanced method in the TSE field and became a powerful tool to convince people who were critical of the prion theory. The transgenic method is used to modify individual genes in animals and plants by directly inserting DNA fragments into cells without using viruses.[4] The method developed rapidly. According to historian of molecular biology Michel Morange, this technique ushered in the age of contemporary molecular biology (Morange, 1998). In the late 1980s, this new wave of molecularization eventually reached the field of scrapie research.

A significant experimental result was reported by one of Prusiner's post-doctoral fellows, Karen Hsiao. She had for some time been interested in Gerstmann–Sträussler–Scheinker (GSS) syndrome, a rare CJD-like human neurodegenerative disease.[5] Hsiao sequenced the PrP gene from GSS cases and found that it carries a mutation, that is, it codes for a variant of the PrP protein (Hsiao and Prusiner, 1990). She observed that one of the DNA sequences in the human prion gene was mutated from proline to leucine at position 102, and speculated that this was the cause of GSS (Hsiao et al., 1989). Hsiao constructed a transgenic mouse containing the GSS prion gene (GSS PrP) that harboured the same mutation. The mutated transgenic mouse eventually died of spongiform neurological disease without prior exposure to scrapie or GSS. In other words, the transgenic mice developed neurodegenerative symptoms spontaneously. This strongly suggested that the disease was indeed caused by a variant form of PrP, which was actually coded by the PrP gene. Prusiner and Hsiao therefore claimed that GSS was an 'inherited prion disease'.

This transgenic experiment made a considerable impression on people both within the scientific community and more widely. A neurobiologist at the Johns Hopkins University School of Medicine, Donald Price, said: 'I think it's really extraordinary. A single mutation in a transgene, when put in a mouse, can cause clinical disease and brain pathology'

(Marx, 1990: 1509). Many Prusiner supporters think that Hsiao's work provided vital data to persuade other scientists that the protein-only hypothesis was accurate.

Again, prion sceptics remained unconvinced. For one thing, they challenged the interpretation that the prion camp had placed on Hsiao's experiment. In particular, they pointed out that the similarities between the neurodegenerative disease suffered by Hsiao's transgenic GSS mice and transmissible diseases such as scrapie were limited. In scrapie, infection was characterized by the presence of a protease-resistant form of PrP. If the pathology of transgenic GSS PrP mice was similar to that caused by scrapie infection, it should be possible to isolate a similar protease-resistant form of PrP from their brains. Indeed, Hsiao and Prusiner tried to do so but failed (Hsiao et al., 1990). Likewise, if GSS was a prion disease, it should be possible to transmit this disease from transgenic GSS PrP mice to ordinary mice. Again, Hsiao and Prusiner attempted it but failed (Hsiao et al., 1990). Richard Carp (IBR) sees in this failure a serious flaw in the evidence, and stresses that 'there has been a whole string of situations, where they transmit material, the Karen Hsiao mouse, where they had the 102 mutation, but if they put it into normal mice they get nothing. So, there has been no instance where artificially produced PrP has been infectious' (Carp, 2000).

Prion sceptics were not satisfied simply to point out gaps in Prusiner's aetiological arguments, however. In 1991, Nora Hunter (NPU) published a paper outlining an alternative theory of the significance of PrP, which, she argued, provided a more adequate account of the theoretical evidence (Hunter, 1991). She claimed that it was not necessary to regard the PrP protein as the infectious agent itself, nor even as a component of the agent. Rather, it could be better explained if it was regarded as a receptor molecule, present in host cells and involved in the process of infection by, and replication of, a scrapie virus. Hunter's view was underpinned by a recent study of the poliovirus receptor (PVR) protein (Ren et al., 1990), which had shown that transgenic mice expressing human poliovirus receptor became susceptible to the poliovirus. Hence, a similar theorization of the PrP protein could account for much of what was known about its involvement in scrapie infection.

At a symposium held in London in 1993, Moira Bruce (NPU) reported that the scrapie-like disease in cattle, BSE, was caused by a single strain of agent, which maintained its identity after passage into other species such as cat, kudu and nyala (Bruce et al., 1994). This implied that the BSE agent behaves independently from the host prion gene. It could also be inferred that the BSE agent has an informational molecule that determines disease characteristics. This paper had a huge impact on many researchers. Chris Bostock, head of the division of molecular biology of the Institute of Animal Health (IAH), says: 'here you have PrP^{Sc} from several different species going into mice and you get the same biological properties. I think the people who support the protein-only hypothesis will find it difficult to explain that' (Kingman, 1993: 181). Even Charles Weissmann acknowledged that 'her [Moira Bruce] results demand a very satisfactory explanation. A very special effort would be needed in order to integrate them into the protein-only hypothesis' (Kingman, 1993: 181)

If no piece of evidence led to the settlement of the controversy, each side nevertheless claims that their own theory has won the war. Stephen DeArmond, a neuropathologist in Prusiner's camp, argues that due to transgenic experiments, the prion theory has gained momentum:

> So, now indirect evidence for the protein-only hypothesis is building more and more. Momentum is gaining. So, now mutations, and then subsequently, all at about the same time, a number of laboratories show that mutations at different points accounted for other forms of CJD as well as other types of GSS-type disorders, and insertions occasionally did it also. So, again, the information is mounting. (DeArmond, 2000)

Moira Bruce at the NPU has a different opinion. She believes that when her team presented their results on the BSE agent at the 1993 symposium,

> it [strain variation] had dramatic effects on people like Charles Weissmann, Stan Prusiner. They just, for some reason, suddenly realised that this has to be explained somehow [...] In that meeting, everybody was talking about the strains, whereas, before this, everybody was just ignoring the whole issue [...] I brought it to the forefront, I think, as a very practical issue, and this approach can be very useful in a practical sense. So that is accepted, in that sense, we won! There is an acceptance that there is an informational component, and 'what it is' is another question. (Bruce, 1999)

Actually, far from coming to an end, the controversy extended to the broader scientific community. In 1994, the Chancellor of the University of California in San Francisco, Joseph Martin, who is also a neurologist, claimed that the prion hypothesis had stood the test of every experiment that could possibly be devised. In contrast, Robert Rowher, director of the molecular neurovirology unit at the Veterans' Affairs Medical Center in Baltimore, urged that the agent is a very hardy and robust virus (Kolata, 1994: C1).

However, the general mood in the scientific community gradually inclined towards acceptance of the protein-only theory, even though every single experimental result could be construed in favour of prion sceptics. As science writer Georgina Ferry wrote in *New Scientist*, 'Prusiner's heresy was to challenge the received wisdom [...] now, more than a decade later, this idea [protein-only theory] is slowly absorbed into mainstream thinking, helping researchers to understand fatal brain diseases (Ferry, 1994: 32).

The triumph of the prion

From the early 1990s, influential agencies within the wider scientific community increasingly came to side with Prusiner and his once heretical suggestion that TSEs are caused by an infectious protein, in the absence of any nucleic-acid-based informational molecule. This growing acceptance of the prion hypothesis manifested itself in the form of numerous significant prizes awarded to Prusiner: a Charles A. Dana Award for Pioneering Achievements in Health in 1992, a Christopher Columbus Quincentennial Discovery Award in Biomedical Research from the National Institutes of Health in 1992, a Gairdner Foundation International Award in 1993, and the Richard Lounsbery Medal from the National Academy of Sciences in 1993 (Spector, 1994: 20). Although his work was not conclusive and still controversial at the time, in 1994 Prusiner was also awarded the Albert Lasker Clinical Medical Research Award, regarded by many as the most significant biomedical science prize in the USA, and generally viewed as a 'predictor' of the Nobel Prize.

Consequently, Prusiner continued to gain support, and in 1997 was awarded the Nobel Prize for physiology or medicine. The Karolinska Institute announced:

> Prusiner has added prions to the list of well known infectious agents including bacteria, viruses, fungi, and parasites. Stanley Prusiner's discovery provides important insights that may furnish the basis of understanding the biological mechanisms underlying other types of dementia-related illnesses – for example, Alzheimer's disease, and established a foundation for drug development and medical treatment strategies. (Karolinska Institutet, 1997)

When the Nobel assembly announced his victory, Prusiner declared: 'concepts are vindicated by the constant actual data and independent verification of data. No prize, not even a Nobel Prize, can make something true that is not true' (Altman, 1997: C3).

The fact is the Nobel announcement was met with hostile criticisms. Many researchers thought that the infectious agent remained unknown, and there was some concern that the Nobel assembly might be prematurely endorsing the controversial theory. Almost all media reports stressed the controversial history of Prusiner's hypothesis, using phrases such as 'once-heretical theory' (Vogel, 1997: 241), 'after years of heated debate' (Coles, 1997: 529), 'controversial research' (Josefson, 1997: 972), and so forth. Laura Manuelidis (Yale) criticized the decision of the Nobel assembly, and claimed that she feared that their endorsement of the prion theory would stifle other avenues of further inquiry. Another prion sceptic, Ashley Haase, a microbiologist at the University of Minnesota, said he thought the Nobel committee should have waited to make the award until there was proof that protein alone was capable of causing infection (Altman, 1997).

The Nobel committee and other general scientists were obviously not disturbed by such criticisms. Ralf Pettersson, deputy chairman of the Nobel committee, claimed that the panel was not bothered by the unanswered questions. The committee was well aware of where the field stood. The details had to be solved in the future. But no one could object to the essential role of the prion protein (Vogel, 1997: 214). Pettersson even suggested that persistent scepticism about the prion theory had contributed to the spread of BSE to humans: 'during the whole of the nineteen-eighties, the prion was very controversial. Acceptance took a while. This could have delayed moves. It was more political decision about when to take action, and by then it was too late' (Rhodes, 1997b: 54–5).

The Nobel Prize seemed to signal the prion's triumph, yet it did not lead to the closure of the prion controversy. In October 1997, just as the Nobel Prize winner was being announced, another series of controversial experimental results came out. Moira Bruce and her team at the NPU published further research on BSE and vCJD. Like their earlier work, this once again raised the issue of strain conservation of the agent upon inter-species passage, as we have seen a phenomenon that could not be explained by Prusiner's prion theory and suggested the existence of an informational molecule (Bruce et al., 1997). Faced with Bruce's results, some commentators claimed that 'whatever the nature of the agent, our understanding of TSE biology is evidently incomplete' (Almond and Pattison, 1997: 438).

Prusiner's Nobel Prize has made the prion hypothesis a mainstream view in the scientific community, especially among disciplinary neighbours. However, while it is true that a great deal of the scientific community accepted Prusiner's theory, the core-set of TSE researchers remained divided.[6] In fact, the majority of this core-set rejected Prusiner's idea. According to Richard Rhodes, at that time only 4 of the 14 major TSE research laboratories working on the infectious agent wholeheartedly espoused the prion theory. Nine others considered it unlikely, and one was undecided (Rhodes, 1997b: 55). Division of opinion within the core-set of TSE researchers is still an issue. For instance, at a recent symposium, another Nobel laureate, Kurt Wüthrich, pointed out that Prusiner's key concept, infectious prion protein, is simply a build-up of garbage (Aguzzi and Heikenwalder, 2003). Despite the fact that Wüthrich contributed to the modelling of a three-dimensional structure of the prion protein, he is still critical of the prion theory, arguing that we must understand the function of the normal prion protein before we can understand prion diseases.[7] In the next section, I will elucidate why Prusiner's research programme was attractive to the scientific community at large, even though the sceptical view was still strong.

Divergent styles of research programme

During the 1970s, Edinburgh researchers gained considerable credibility in the scientific community because, among other things, they had a broadly biological perspective that was in keeping with the approaches favoured by the British research councils (Kim, 2000). This perspective remained the predominant approach to scrapie and related diseases in Britain as well as in the USA through the 1980s. Prusiner's early work started to deviate from mainstream scrapie research in that it abandoned the broadly biological approach for a much narrower focus on the biochemical aspects of the agent. Each camp has since developed and maintained distinctive intellectual and methodological frameworks in the course of the dispute. The controversy was actually generated and sustained by the confrontation of these two patterned styles of research programme.

As we have seen, the two groups of scientists produced various experimental results, and each group believed that their results supported their respective theory on the nature of the scrapie agent. However, the experimental evidence failed either to decide unequivocally between the two sets of theories, or to establish common ground on which the two groups could have reached a consensus. This was because the two groups evaluated the data with quite different criteria, themselves embedded in two different research programmes: on the one hand, a *generalist biological* programme and on the other, a *specialist biochemist* programme. Prion sceptics have concentrated on the complicated phenomena of *disease*. Their main concern was how the disease replicates in the host, and how the agent and the host genes interact. In contrast, the prion group has mainly focused on the biochemical structure of the *agent*. The different primary goals and research orientations led the two groups to construct quite distinct experimental programmes. Prion sceptics' concentration on the nature of the disease led them to explore a variety of biological phenomena, including the nature of the agent. Prusiner's concentration on the nature of the agent led him only secondarily to ask how it is implicated in the disease. The two factions thus pursued strikingly distinct ranges of intellectual and methodological issues.

Generalist programme of prion sceptics

Prion sceptics have mainly focused on disease aetiology, pathogenesis and agent–host interactions as well as on the nature of the infectious agent. Indeed, this biological and pathological orientation was clear in the articles that opened the prion controversy in 1982 (Kimberlin, 1982a, b). More specifically, for Dickinson and colleagues, the biological diversity of the scrapie agent provided the main point of attack for elucidating its chemical nature. Since the 1960s, they have identified several strains of scrapie. They have also realized that strain variation poses a major challenge to the prion theory. For this reason, Dickinson and colleagues stress the importance of understanding the nature of agent variation. As Dickinson commented at the UK BSE Inquiry, this distinguished his own approach from that adopted by Prusiner:

> The NPU had distinguished itself from most work worldwide, when most people were saying: 'we want to know what the nature of this agent is'. And I, starting as a geneticist, said: 'I think a more fundamental question is: "what is the nature of agent variation?"' [...] It is very important distinction. If you think about it, there are those who claim, I think prematurely, that they know what the nature of the agent is in chemical terms. The outstanding question is very much: 'what are strain differences?' 'what is the nature of agent variation?' (Dickinson, 1999a: 4–5)

In this respect, Edinburgh researchers developed a distinctive intellectual and experimental programme. Interestingly, they refer to their overall experimental project as a 'generalist project'. As George Outram remarks, 'scientists themselves fall into two kinds of fundamental types generally, generalist and specialist ... our culture, I can say, is generalist. You've got a scientist who knows a lot and is very good at some techniques and extremely complicated equipment. Then you have generalist' (Outram, 1999). In this statement, Outram does not only identify himself as a generalist, but also points to the generalist orientation of the entire research programme that he and his Edinburgh colleagues developed. Earlier, Outram had spelled out the philosophical grounds of their research project:

> The danger with this approach [biochemistry, immunology, virology and molecular biology apply here] is that in order to get meaningful answers the right questions must first be asked and, if scrapie is an unprecedented phenomenon, then the inbuilt assumptions of any developed methodology will effectively prevent the agent from 'answering' the questions we address to it. In short, we require something more *general*, i.e., less specialised, which will survey the whole phenomenon and so enable us to identify or devise such specialised techniques as will be really appropriate. (Outram, 1980: 360)

Edinburgh researchers thus concluded that before exploring the specific characteristics of the agent, their research should aim at providing a general understanding of the disease: 'this should provide a broad biological base against which the disease could declare itself in its own terms rather than those imposed by some other inappropriate system' (Outram, 1980: 360). For Edinburgh researchers, strain variation was the best subject to examine, because an understanding of strain variation would throw light on the whole biological mechanism of the disease.

This distinctive intellectual orientation was also embodied in their research methodology. They have studied as many aspects of strain variation as possible, including incubation time, lesion profiles and the effects of host genotype on both of these. To this end, they needed an animal model that would make it possible both to display the widest range of strains possible, and to standardize the biological circumstances in which those strains were investigated. In this respect, it is notable that after succeeding in transmitting scrapie to hamsters and demonstrating that the disease incubated more quickly than in mice, Dickinson did *not* adopt the hamster as his preferred experimental animal. The reason is simple: whereas at least 20 different strains of scrapie could be studied in mice, only two of these could be transmitted to hamsters. Thus, if hamster experiments could be performed much faster, they did not display the range of phenomena that Edinburgh researchers considered imperative to observe (Kimberlin and Walker, 1977).

Specialist programme of the prion group

In contrast to prion sceptics' general biological perspective, Prusiner's work concentrated primarily on the agent, while aetiological and pathological phenomena were secondary to him. On the whole, he was interested less in how the agent manifests itself in the *form of disease*, than in simply asking what the *agent is*, particularly in chemical and molecular biological terms. He clearly set out his priorities in his 1982 reply to the criticisms expressed by Dickinson and Kimberlin: 'Knowledge of the structure of the agent is mandatory before attempting to design studies that can answer such fundamental and critical questions as: (1) how does the agent replicate, (2) in what cells does it replicate, and (3) how does it produce neurological dysfunction?' (Prusiner, 1982b: 494).

Prusiner's view of the objective of his research was thus the precise opposite of what Outram suggested. Conversely, Prusiner and his team were able to bring a range of specialist methods to bear on their research. As one of Prusiner's colleagues, Mike Scott, remarks: 'We're all becoming specialist in some way. I think you have to go with technology, and the technology is molecular biology, rational drug-design, genomics, NMR, X-ray crystallography, recombinant antibodies etc... I think anybody who doesn't embrace the new "specialist" biotechnology is doing himself or herself a disservice' (Scott, 2000).

For Prusiner and his team, understanding the molecular structure of the agent by using various techniques of biochemistry was the first step towards exploring the whole biological mechanism of the disease. Since he launched his research project to isolate the agent from the cellular components of infected material in the 1970s, Prusiner has consistently adopted some of the most innovative biochemical and molecular biological techniques in a sustained assault on this single problem.

In this context, Prusiner's preferred choice of experimental animal also distinguished him from the sceptics. As we have seen, the Edinburgh group favoured mice because they revealed a wide range of scrapie phenomena. Prusiner, in contrast, was initially interested in laboratory animals only as a means of performing bioassays of the scrapie agent after exposure to various physical and chemical treatments. For this purpose, the hamster with its shorter incubation period was far

Table 2.1 Stylistic differences between the prion advocates and prion sceptics

	Generalist biological style	*Specialist biochemist style*
Infectious agent	Small putative nucleic acids	Protease-resistant proteins (PrP)
Research goal	Nature of biological diversity	Biochemical nature of the agent
Experimental model	Mouse	Hamster
Titration method	Conventional end-point method	Incubation time method
Experimental techniques	Conventional genetics; pathological examination; electron microscopy	Chemical treatments; transgenic mice; NMR, etc.
Existence of genome	Yes	No
Hypothetical agent	Virino	Prion

better than mice. Since 1989, Prusiner's team has constructed transgenic mice models as a means of focusing yet more closely on the proteinaceous character of the scrapie agent. At each stage of his experimental project, he has adapted animals more as a means to apply particular biochemical techniques than for the biological phenomena they reveal. In this sense, Prusiner's experimental system is more specialist-oriented than that of the Edinburgh group.[8]

Table 2.1 provides a synthetic view of the differences between the two experimental styles described above.

Prions and the molecularization of biomedicine

Molecularizing prions

As we have seen, Prusiner lost the support of the scrapie research community in the early stages of the prion controversy. Though besieged by sceptics, however, he escaped the siege by collaborating with scientists outside the scrapie community. Since then, he has tended to address scientists in neighbouring disciplines as well as the members of the immediate scrapie research community. He has been remarkably successful in attracting collaborators from outside the TSE field, and has also succeeded in publicizing his work and in convincing many that it is valuable. This has been decisive in winning him the widespread support that his ideas now enjoy in the scientific community as a whole.

This raises the question of why disciplinary neighbours have inclined to the prion theory. I would argue that Prusiner's style of scientific programme is in keeping with the way many areas of biomedical science have been developing in recent years. Molecular biological approaches, in particular, are very closely linked to the commercialization of contemporary biomedicine. Many molecular biologists are becoming increasingly influential within science – not least because they are becoming increasingly rich and increasingly closely connected to other powerful institutions in industry and government. Prusiner has been dealing with many of those people – hiring or collaborating with them, buying or selling scientific products, and so on. Indeed, his own style of practice epitomizes the values of this market-oriented new scientific culture, while his ability to claim, if not to prove, fundamental breakthroughs in understanding TSEs serves to vindicate the whole enterprise.

During the 1980s and 1990s, the application of various new techniques accelarated the so-called molecularization of biology. Joan Fujimura has described how with the commitment of one research group after another to the pursuit of proto-oncogenes in cancer research, molecular biological research has become a bandwagon (Fujimura, 1988). She later argued: 'Large numbers of people, laboratories, organisations, and resources became committed to one approach to a problem' (Fujimura, 1996: 2). A crucial element in this convergence of scientific interests was the availability and adoption of standardized technologies such as NMR (nuclear magnetic resonance), PCR (polymerase chain reaction), transgenic technique, and so forth. Indeed, as Steve Sturdy argues, standardization was a vital element in the growth of molecularization more generally, and facilitated the exchange of data, materials, and ideas within what might be called a 'molecular economy' (Sturdy, 1998).

Prusiner hitched his own scrapie research to the molecular biological bandwagon in the mid-1980s, when he began to look for the PrP gene. By 1989, when he began to manufacture transgenic mice, he was a keen proponent of molecular biological methods. DeArmond suggests that these transgenic experiments played a significant role in winning him the support of fellow scientists. He argues that with the transgenic model, Prusiner's group gained momentum (DeArmond, 2000). Just after reporting Prusiner's success in transgenic experiments, the *New York Times* claimed that the once heretical theory had now gained huge credibility within the scientific community (Blakeslee, 1991: C12). This is confirmed by the increasing usage of prion terminology in the TSE field, particularly aggressive between 1990 and 1992. This coincides with the publication of results of the transgenic experiments conducted by Prusiner and colleagues between 1989 and 1992. In fact, before the transgenic experiments, Prusiner's position in the scientific community was still defensive, and many regarded his theory as heretical. By 1992, however, the map of the controversy had been redrawn, and the general mood of the scientific community had come down on the side of the prion theory. As such transgenic technology played a pivotal role in persuading fellow scientists.

Several reasons explain why Prusiner's participation in the molecular bandwagon won him the support of his fellow scientists. In the first place, his use of transgenic mice aligned his scrapie investigations with what was widely regarded as cutting-edge developments in other areas of biomedicine. Prusiner's application of new molecular biological techniques to the elucidation of a puzzling and important family of diseases vindicated the investment that had been made in molecular biology by scientists from a wide range of biomedical disciplines. Consequently, many enthusiastic partisans of molecular biological and especially transgenic techniques as standardized research tools in bioscience, ranked among those who praised Prusiner's transgenic mouse work (Sofroniew and Staley, 1991; Hardy, 1991).

In turn, scientists working in other fields such as Alzheimer's disease attempted to apply Prusiner's idea about post-translational change of protein to their own research. In part, this was presumably because Prusiner's innovative use of exciting new molecular biological techniques appeared to offer a possible way out of the stagnating programme of research into unconventional slow viruses and other supposedly similar neurodegenerative disorders. The appeal of Prusiner's transgenic method also stemmed from the possibility that it might open the way to a possible pharmaceutical solution for these diseases. It engaged directly with the interests of the pharmaceutical and biotech companies. Although so-called prion diseases are generally rare in humans, the pharmaceutical issue became more significant with the bovine spongiform encephalopathy (BSE) crisis in Britain. In 1996, scientists claimed that the new variant of Creutzfeldt–Jakob disease (vCJD) identified in the UK represented BSE transmission to humans. The scientific community thus urgently needed to produce effective means of stopping this potential epidemic of vCJD. Prusiner's transgenic work presaged a way to satisfy this requirement in pointing to the possible development of molecular biological techniques of diagnosis and treatment.

The pharmaceutical and biotechnology industries came to see that their interests were involved. Since then, prion researchers have rushed to set up biotechnology companies to commercialize new methods of diagnosis and possible clinical treatments. In 1997, some of Charles

Weissmann's colleagues established a biotechnological company called 'Prionics Inc.', which has already commercialized two diagnostic methods: Prionics®-Check WESTERN and Prionics®-Check LIA. At least 35 laboratories are currently searching for commercial diagnostic and clinical possibilities in this field. Moreover, Prusiner's team attempted the first pharmacotherapeutic treatment of vCJD, though it was unsuccessful.[9] The prion group's adoption of standard biotechnological methods thus met the needs and expectations of wider groups in the scientific community, including the biotech industry. As a result, it strengthened the position of the prion group.

Prion sceptics, on the other hand, especially Edinburgh researchers, continued to focus on classical genetic and pathological techniques such as measuring incubation period and pathological change. These techniques were intended to engage the attention of the immediate scrapie scientific community, and played a part in building a consensus during the 1960s and 1970s. However, once the molecularization of biology came into play, such techniques were no longer fashionable and less likely to be the object of attention of the scientific community. At times, the lack of a molecular approach at the NPU was seriously considered by the members of the institute, and attempts were made to recruit molecular biologists such as Jean Manson. It remains that before 1986 the NPU did not pursue any biochemical or molecular biological projects. When Prusiner and his team produced significant experimental data on protein sequencing and found the prion gene in 1985, the NPU failed to respond.

Linkage of prion research to wider social networks

For 20 years, the opposing scientific programmes of prion advocates and prion sceptics have been competing. However, equipoise in the controversy has gradually broken down, and the specialist programme of the prion group has gained the upper hand. This does not mean that Prusiner's programme has an innate superiority. As we have seen, the two factions produced many valuable experimental data, but failed to reach an agreement. Hence, the prion controversy cannot be said to have reached closure. Rather, the voices of prion sceptics, though insistent, simply attract less attention than hitherto. Consequently, there is no point in attempting to explain this shift in opinion by pointing to some definitive piece of empirical research or theoretical insight. Rather, we must look at the shifting networks of interconnected scientific interests that have constructed the credibility of the prion group within the wider scientific community and beyond.

In the 1960s, the small size of the scrapie research community enabled a handful of researchers in Edinburgh to establish a coherent agenda for collaborative research. Thus, the primary task at the time was consolidation of the research network rather than expansion of the community. Based on generally accepted genetic and pathological methods, the generalist programme of Dickinson and colleagues became mainstream in scrapie research. In addition, the failure of biochemical and biophysical work on scrapie shaped a general feeling of frustration with such approaches.[10] Hence, during the 1970s the generalist biological approach was the dominant one in scrapie research, and in so far as the wider scientific community took an interest in it, it tended to regard the pathogenic studies of Edinburgh as definitive (Kim, 2000).

It should be stressed that the Edinburgh research programme was initially funded by a research council, the Agricultural Research Council (ARC, later Agricultural and Food Research Council). Under such funding structure researchers have a stable post. As Moira Bruce says: 'everybody has a permanent job. There is no problem about the short-term contracts. It was stabilised funding. It was fairly open-ended and remarkably relaxed' (Bruce, 1999). Crucially, with this stable funding structure it was possible to conduct the long-term experiments entailed by the long incubation period of scrapie in mice.

The genetic and pathology projects at the NPU were conditional upon stable and uninterrupted funding to guarantee continuity of work. However, such funding and the long-term collaborative research it allowed became less and less the norm in the scientific community, even within the British system of research council funding. At the NPU, the imposition of greater managerial controls over scientists, and pressure for more commercially relevant lines of research, led Dickinson to resign as director in 1987, over fears that the autonomy of his unit was under threat (Biggs, 1998: 58–60). Genetically oriented research on scrapie and BSE continued under Moira Bruce and others, but it was increasingly integrated into a wider programme of research on animal diseases that tended to favour biochemical and molecular biological approaches (Institute of Animal Health, 1998: 3–4).

Meanwhile, in the USA governmental science policy was changing even more markedly in favour of the commercialization of biomedicine. In the 1980s, the American Congress voted laws that allowed personal profit to be made from university or federal research achievements supported by federal funds.[11] The US government thus changed their basic idea on public ownership of scientific research. Thereafter, commercialization of the life sciences accelerated. Commercial molecular biology was at the forefront of this process. Interestingly, one of Prusiner's key collaborators, Charles Weissmann, is a frontrunner in the commercialization of molecular biology. He was a founding member of the first biotech company in Europe, Biogen, in the late 1970s (Wright, 1994: 87). Another of his collaborators, Leroy Hood, also established a biotech company, the Institute for Systems Biology, in 1999.

Prusiner's market-oriented and contractual style of practice thus corresponds closely to the kind of scientific activity favoured by policy makers and is seen to support some of the most commercially exciting developments in modern biomedicine. In a context of molecularization, which includes commercialization and standardization of biomedical techniques, increasingly large sections of the scientific community tend to favour the values of the socio-economic world that Prusiner inhabits – a world of short-term contracts, deals and patents, and so on. Though prion sceptics may still be doing valuable work, they are largely isolated from this new scientific culture. It is for this reason that their generalist biological programme has become less convincing and less interesting than the specialist molecular programme pursued by Prusiner. Prusiner's triumph in the prion controversy is ultimately due not to any inherent superiority of his theories or data, but to the peculiar scientific culture that tends to favour his entire way of doing science over that of his opponents.

Conclusion

In this chapter, I have described the prion controversy that opposed prion supporters and prion sceptics during the 1980s and 1990s. I have addressed two major questions. Firstly, why the controversy emerged and persisted for 20 years. Secondly, why the prion group eventually gained much support from the scientific community, as demonstrated by the Nobel Prize awarded to Prusiner.

I have shown that the controversy was rooted in the opposition of two distinct styles of research programme. The fundamental disagreement between prion advocates and prion sceptics on the interpretation of experimental results derived from their divergent research priorities and methodologies, which I have characterized in terms of a generalist biological programme on the one hand, and a specialist biochemist programme on the other. Prion sceptics adopted the former style, whereas the prion group adopted the latter. These divergent styles in turn embodied different aims, methodologies and experimental systems of research, which prevented the emergence of a consensus. Unsurprisingly, the two factions of researchers have maintained their respective research programmes during the 20-year controversy and are still disagreeing.

I have also shown that the apparent victory of the prion group cannot be explained by the inherent superiority of their theoretical or experimental work. The prion group's style of practice conforms more closely than that of prion sceptics to the developments that have recently been taking place in biomedical science. Modern biomedicine tends increasingly towards commercialization and standardization, including individual competitions, short-term contracts, and the use of standardized tools of research. This large-scale shift in biomedicine is often defined in terms of 'molecularization'. The prion group is located at the forefront of this shift, whereas the sceptics are largely isolated from it. Prusiner's programme is thus intimately linked to the wider social networks that are influential in determining how scientific funding and credit are distributed. The dominance of his prion theory is therefore the outcome of the social and cultural position he occupies, and of his ability as a scientist to fulfil the expectations that such a position brings with it.

Acknowledgements

I would like to thank Dr Steve Sturdy and Professor David Bloor for reading and commenting on this chapter. I would also like to express my gratitude to Dr Eve Seguin for reading it and making valuable comments. Finally I would like to thank Carole Tansley for correcting my English. This chapter is based on my PhD thesis, undertaken in the Science Studies Unit at the University of Edinburgh.

Notes

1. The Neuropathogenesis Unit (NPU) was established in 1981 with funds provided by the Agricultural Research Council (ARC, which the same year became the Agricultural and Food Research Council, AFRC) and by the Medical Research Council (MRC). In 1978, the ARC's Advisory Committee on Scrapie decided that research institutes working on scrapie, i.e. the joint scrapie research unit of the Moredun Institute and Animal Breeding Research Organization (ABRO) headed by Alan Dickinson in Edinburgh, and the Institute for Research on Animal Diseases (IRAD) in Compton, should be transferred to one location in order to set up a new TSE research centre. The committee decided to set it up in Edinburgh. Consequently, the scrapie research programme at IRAD was terminated, and its resources were transferred to Edinburgh. With the exception of Richard Kimberlin who had been collaborating with Edinburgh researchers since the mid-1970s, IRAD researchers were

excluded in this restructuration process. As a result, the NPU mostly consisted of the former members of the Moredun–ABRO unit: Alan Dickinson, George Outram, Hugh Fraser and Moira Bruce. For a more detailed analysis of this restructuring process, see Kim (2000).

2. In an interview Dickinson admitted to being the anonymous author of the *Lancet* editorial (Dickinson, 1999b).
3. Dickinson and his group at the NPU in Edinburgh, Heino Diringer's group at the Robert Koch Institute in Berlin, Laura Manuelidis' group at the Yale Medical School, Richard Carp's group at the Institute of Basic Research (IBR) in New York, Carleton Gajdusek's group at the National Institute of Neurological Disorders and Stroke (NINDS) in Bethesda, William Hadlow's group at the Rocky Mountain Laboratory (RML) in Hamilton, Richard Marsh's group at the Veterinary School of the University of Wisconsin in Madison, and Prusiner's group at the University of California in San Francisco (UCSF).
4. This method was devised by two researchers, F. Graham and A. Van Der Erb, in 1973 (Graham and Van der Erb, 1973). The main progress of the transgenic technique occurred when transgenic animals were created in 1980. A Yale biologist, Frank Ruddle, injected mouse embryos a few hours old with foreign DNA that then integrated into their chromosomes (Gordon et al., 1980). After several rounds of cell division *in vitro*, the embryos were implanted into surrogate mothers, which, 20 days later, gave birth to a total of 78 baby mice, two of which had integrated the foreign DNA into most of their cells (Morange, 1998: 202).
5. GSS is one of the human TSEs. In 1936, two neurologists, Gerstmann and Sträussler, and a neuropathologist, Scheinker, described a family with unusual neurodegenerative symptoms. It is an extremely rare disease which strikes only one in about 10 to 100 million people. Some of its pathological features resemble those of Alzheimer's disease (Masters et al., 1981).
6. The concept of core-set was first suggested by Harry Collins. In his work on gravity waves, he defines core-set as follows: firstly the core-set of scientists are those who are actively involved in experimentation or observation; secondly, they are making contributions to the theory of the phenomenon or of the experiment. Core-set provides us with a good concept to account for this divisive phenomenon in the prion controversy. For more detail, see Collins (1981).
7. Wüthrich won the Nobel Prize for his development of nuclear magnetic resonance spectroscopy for determining the three-dimensional structure of biological macromolecules in solution. Furthermore, around 1996, he applied his NMR technique to reveal the three-dimensional structure of the prion protein. For a more detailed description of his contribution to prion research, see Segal and Francoeur, Chapter 4 this volume.
8. It should be noted that the demarcation between generalist and specialist programmes should not be perceived as an absolute distinction. At some level of organization and collaboration, this divergence is at most a matter of degree, not one of kind. For instance, Dickinson admits that the Edinburgh team's progress was to an extent delayed by the lack of biochemists, which shows that they recognize and rely on specialist skills. Thus, the demarcation between generalist and specialist programmes points to distinct tendencies in terms of research objectives, methodologies and experimental systems that are best captured by the term 'style'.

9. This is the famous case of vCJD victim Rachel Fober who went to San Francisco for clinical treatment led by Prusiner and Korth. It had some success but the victim eventually died (Dealler, 2001; Brockes, 2002; Korth et al., 2001).
10. In the 1960s and 1970s, a large biochemical research programme on scrapie existed at the IRAD in Compton, and collaboration with Tikvah Alper, a radiobiologist at Hammersmith Hospital in London, produced some interesting findings (Alper et al., 1967). Unfortunately, IRAD researchers failed to identify the basic biochemical characteristics of the scrapie agent. In addition, the programme was undermined by a series of experimental fiascos (Pattison and Jones, 1968; Hunter, 1992). As a result, many scrapie researchers came to regard the IRAD's biochemical programme as unproductive.
11. The laws are the Bayh–Dole Act and the Stevenson–Wydler Act in 1980, and the Federal Technology Transfer Act in 1986 (Andrews and Nelkin, 2001).

References

Note: Interview references are detailed separately after published references.

Aguzzi, A. and M. Heikenwalder (2003) 'Cannibals and Garbage Piles', *Nature* 423: 127–9.

Almond, J. and J. Pattison (1997) 'Protein Only Prions?', *Nature* 389: 438.

Alper, T. et al. (1967) 'Does the Agent of Scrapie Replicate without Nucleic Acid?', *Nature* 214: 764–6.

Altman, L. K. (1997) 'US Scientist wins Nobel Prize for Controversial Work', *The New York Times* 7 October: C3.

Andrews, L. and D. Nelkin (2001) *Body Bazaar: the Market for Human Tissue in the Biotechnology Age*. New York: Crown Publishers.

Anonymous Editorial [Alan Dickinson] (1982) 'Scrapie: Strategies, Stalemates, and Successes', *Lancet* 1: 1221–3.

Basler, K. et al. (1986) 'Scrapie and Cellular PrP Isoforms are Encoded by the Same Chromosomal Gene', *Cell* 46: 417–28.

Biggs, P. (1998) 'Transcription of Oral Hearings: Day 44', *The BSE Inquiry* 8 July.

Blakeslee, S. (1991) 'Heretic Theory on Brain Diseases Gains New Ground', *New York Times* 8 October: C1.

Bolton, D. C. et al. (1982) 'Identification of a Protein that Purifies with the Scrapie Prion', *Science* 218 (4579): 1309–11.

Brockes, E. (2002) 'To the Last Breath', *Guardian* 15 January G2: 1–2.

Brown, P. (1999) 'Charles Weissmann: Another New Challenge', *Current Biology* 9 (17): R625.

Bruce, M. E. and H. Fraser (1991) 'Scrapie Strain Variation and its Implications', in B. W. Chesebro (ed.) *Current Topics in Microbiology and Immunology 172: Transmissible Spongiform Encephalopathies – Scrapie, BSE and Related Disorders*. London: Springer Verlag: 125–38.

Bruce, M. et al. (1994) 'Transmission of Bovine Spongiform Encephalopathy and Scrapie to Mice: Strain Variation and the Species Barrier', *Philosophical Transactions of the Royal Society, London, Series B, Biological Series* 343: 405–11.

Bruce, M. et al. (1997) 'Transmissions to Mice Indicate that "New Variant" CJD is Caused by the BSE Agent', *Nature* 389: 498–501.

Chesebro, B. et al. (1985) 'Identification of Scrapie Prion Protein-Specific mRNA in Scrapie-infected and Uninfected Brain', *Nature* 315: 331–3.

Coles, H. (1997) 'Nobel Panel Rewards Prion Theory after Years of Heated Debate', *Nature* 389: 529.

Collins, H. M. (1981) 'The Place of the Core-set in Modern Science: Social Contingency with Methodological Propriety in Science', *History of Science* 14: 6–19.

Cooke, J. (1998) *Cannibals, Cows and the CJD Catastrophe*. Sydney: Random House.

Crick, F. (1970) 'Central Dogma of Molecular Biology', *Nature* 227: 561–3.

Dealler, S. (1996) *Lethal Legacy: BSE–Search for the Truth*. London: Bloomsbury.

Dealler, S. (2001) 'At Long Last, Signs of a BSE Breakthrough', *Guardian* 5 September: 16.

Dickinson, A. G. (1976) 'Scrapie in Sheep and Goats', in R. H. Kimberlin (ed.) *Slow Virus Disease of Animal and Man*. Amsterdam: North-Holland Publishing Co.: 209–41.

Dickinson, A. G. (1999a) 'Transcription of Oral Hearings: Day 31', *The BSE Inquiry* 11 June.

Dickinson, A. G. and V. M. Meikle (1969) 'A Comparison of Some Biological Characteristics of the Mouse-Passaged Scrapie Agents, 22A and ME7', *Genetical Research* 13 (2): 213–25.

Dickinson, A. G. and G. W. Outram (1979) 'The Scrapie Replication-site Hypothesis and its Implications for Pathogenesis', in S. B. Prusiner and W. J. Hadlow (eds) *Slow Transmissible Diseases of the Nervous System* 2. London: Academic Press: 13–31.

Dickinson, A. G. et al. (1968) 'Identification of a Gene which Controls the Incubation Period of Some Strains of Scrapie Agent in Mice', *Journal of Comparative Pathology* 78 (3): 293–9.

Diringer, H. et al. (1983) 'Scrapie Infectivity, Fibrils and Low Molecular Weight Protein', *Nature* 306: 476–8.

Ferry, G. (1994) 'Mad Brains and the Prion Heresy', *New Scientist* 28 May: 32–7.

Fujimura, J. H. (1988) 'The Molecular Biological Bandwagon in Cancer Research: Where Social Worlds Meet', *Social Problems* 35 (3): 261–83.

Fujimura, J. H. (1996) *Crafting Science: a Sociohistory of the Quest for the Genetics of Cancer*. Cambridge, Mass.: Harvard University Press.

Fujimura, J. H. and D. Y. Chou (1994) 'Dissent in Science: Styles of Scientific Practice and the Controversy over the Cause of AIDS', *Social Science and Medicine* 38 (8): 1017–36.

Fujimura, J. H. and D. Y. Chou (1995) 'Styles of Practice in HIV/AIDS Research', *Techniques and Culture* 25–6: 195–244.

Gordon, J. W. et al. (1980) 'Generic Transformation of Mouse Embryos by Microinjection of Purified DNA', *PNAS* 77: 7380–4.

Graham, F. L. and A. J. Van der Erb (1973) 'A New Technique for the Assay of Infectivity of Human Adenovirus DNA', *Virology* 52: 456–67.

Hardy, J. (1991) 'Prion Dimers: a Deadly Duo?', *Trends in Neuroscience* 14 (10): 423–4.

Harwood, J. (1987) 'National Styles of Science: Genetics in Germany and the United States between the World Wars', *ISIS* 78: 390–414.

Harwood, J. (1993) *Styles of Scientific Thought: the German Genetics Community 1900–1933*. Chicago: University of Chicago Press.

Hsiao, K. K. and S. B. Prusiner (1990) 'Inherited Human Prion Diseases', *Neurology* 40 (12): 1820–7.
Hsiao, K. K. et al. (1989) 'Linkage of the Prion Protein Missense Variant to Gersmann–Straussler Syndrome', *Nature* 338 (6213): 342–5.
Hsiao, K. K. et al. (1990) 'Spontaneous Neurodegeneration in Transgenic Mice with Mutant Prion Protein', *Science* 250 (4987): 1587–90.
Hunter, G. D. (1992) 'The Search for the Scrapie Agent', in S. B. Prusiner et al. (eds) *Prion Diseases of Humans and Animals*. London: Ellis Horwood: 23–9.
Hunter, N. (1991) 'Scrapie and GSS – the Importance of Protein', *Trends in Neurosciences* 14 (9): 389–90.
Institute of Animal Health (1998) 'Research on Transmissible Spongiform Encephalopathies, 1986–1998: an Overview', *The BSE Inquiry: statement 105.*
Josefson, D. (1997) 'The Prion Hypothesis is Finally Accepted by the Establishment', *British Medical Journal* 315: 972.
Karolinska Institutet (1997) 'The Nobel Assembly at the Karolinska Institute has Today Decided to Award the Nobel Prize in Physiology or Medicine for 1997 to Stanley B. Prusiner', *Press Release* (6 October).
Kim, K. (2000) 'Controversy on the Nature of the Scrapie Agent in the 1960s', paper delivered at the Fourth British–North American Joint Meeting of the History of Science, St Louis, USA: 3–6 August.
Kimberlin, R. H. (1979) 'Early Events in the Pathogenesis of Scrapie in Mice: Biological and Biochemical Studies', in S. B. Prusiner and W. J. Hadlow (eds) *Slow Transmissible Diseases of the Nervous System* Vol. 2. New York: Academic Press: 33–54.
Kimberlin, R. H. (1982a) 'Reflections on the Nature of Scrapie Agent', *Trends in Biochemical Sciences* 7: 392–4.
Kimberlin, R. H. (1982b) 'Scrapie Agent: Prions or Virinos?', *Nature* 297: 107–8.
Kimberlin, R. H. and C. A. Walker (1977) 'Characteristics of a Short Incubation Model of Scrapie in the Golden Hamster', *Journal of General Virology* 34: 295–304.
Kimberlin, R. H. et al. (1989) 'The Genomic Identity of Different Strains of Mouse Scrapie is Expressed in Hamsters and Preserved on Reisolation in Mice', *Journal of General Virology* 70: 2017–25.
Kingman, S. (1993) 'London Meeting Explores the Ins and Outs of Prions', *Science* 262: 180–1.
Kolata, G. (1994) 'Viruses or Prions, an Old Medical Debate Still Rages', *New York Times* 4 October: C1.
Korth, C. et al. (2001) 'Acridine and Phenothiazine Derivatives as Pharmacotherapeutics for Prion Disease', *PNAS* 98 (17): 9836–41.
Lane, N. F. (1997) 'Double Helixes and Double-edged Swords: Cloning and the Conundrum of Scientific Success', *National Press Club* 22 April: www.npf.gov.od/lpa/forum/lane/n1497/npc.htm
Lax, A. J. et al. (1983) 'Involvement of Protein in Scrapie Agent Infectivity', *Research in Veterinary Science* 34: 155–8.
Mannheim, K. (1953) 'Conservative Thought', in *Essays in Sociology and Social Psychology*. London: Routledge and Kegan Paul: 74–164.
Manuelidis, L. et al. (1985) 'Specific Proteins Associated with Creutzfeldt–Jakob Disease and Scrapie Share Antigenic and Carbohydrate Determinants', *PNAS* 82: 4263–7.

Manuelidis, L. et al. (1987) 'Evidence Suggesting that PrP is not the Infectious Agent in Creutzfeldt–Jakob Disease', *EMBO Journal* 6 (2): 341–7.

Marx, J. (1990) 'Human Brain Disease Recreated in Mice', *Science* 250 (4987): 1509.

Masters, C. L. et al. (1981) 'Creutzfeldt–Jacob Disease Virus Isolation from the Gerstmann–Straussler Syndrome with an Analysis of Various Forms of Amyloid Plaque Deposition in the Virus-induced Spongiform Encephalopathies', *Brain* 104: 559–88.

Millson, G. C. et al. (1976) 'The Physico-chemical Nature of the Scrapie Agent', in R. H. Kimberlin (ed.) *Slow Virus Diseases of Animals and Man*. Oxford: North-Holland Publishing Co.: 243–66.

Morange, M. (1998) *A History of Molecular Biology*. Cambridge: Harvard University Press.

Oesch, B. et al. (1985) 'A Cellular Gene Encodes Scrapie PrP 27–30 Protein', *Cell* 40: 735–46.

Outram, G. W. (1976) 'The Pathogenesis of Scrapie in Mice', in R. H. Kimberlin (ed.) *Slow Virus Diseases of Animals and Man*. Amsterdam: North-Holland Publishing Co.: 325–57.

Outram, G. W. (1980) 'Mouse Scrapie: Black-box Models and the Slow Encephalopathies', in F. C. Rose and P. O. Behan (eds) *Animal Models of Neurological Disease*. Bath: Pitman Medical Limited: 357–77.

Pattison, I. H. and G. C. Millson (1961) 'Scrapie Produced Experimentally in Goats with Special Reference to the Clinical Syndrome', *Journal of Comparative Pathology* 71: 101–8.

Pattison, I. H. and K. M. Jones (1968) 'Detection of the Scrapie Agent in Tissues of Normal Mice and in Tumours of Tumour-bearing but Otherwise Normal Mice', *Nature* 217 (136): 102–4.

Perlman, D. (1982) 'Tiny Life Form Found', *San Francisco Chronicle* 19 February: 1.

Prusiner, S. B. (1982a) 'Novel Proteinaceous Infectious Particles Cause Scrapie', *Science* 216: 136–44.

Prusiner, S. B. (1982b) 'Research on Scrapie [Letter]', *Lancet* 2 (8296): 494–5.

Prusiner, S. B. (1984) 'Prions: Novel Infectious Pathogens', *Advances in Virus Research* 29: 1–56.

Prusiner, S. B. et al. (1984) 'Purification and Structural Studies of a Major Scrapie Prion Protein', *Cell* 38: 127–34.

Ren, R. et al. (1990) 'Transgenic Mice Expressing a Human Poliovirus Receptor: a New Model for Poliomyelitis', *Cell* 63 (2): 353–62.

Rhodes, R. (1997a) *Deadly Feasts*. New York: Simon and Schuster.

Rhodes, R. (1997b) 'Pathological Science', *New Yorker* 1 December: 54–9.

Sklaviadis, T. K. et al. (1989) 'Physical Properties of the Creutzfeldt–Jakob Disease Agent', *Journal of Virology* 63 (3): 1212–22.

Sofroniew, M. V. and K. Staley (1991) 'Transgenic Modelling of Neurodegenerative Events Gathers Momentum', *Trends in Neuroscience* 14 (12): 513–14.

Spector, B. (1994) 'Lasker Awards Cite Persistence of Three Scientists', *The Scientist* 8 (17 October): 20.

Sturdy, S. (1998) 'Reflections: Molecularisation, Standardization, and the History of Science', in S. de Chadarevian and H. Kamminga (eds) *Molecularising Biology and Medicine: New Practices and Alliances, 1910s–1970s*. Amsterdam: Harwood Academic Publishers: 273–92.

Taubes, G. (1986) 'The Game of the Name is Fame. But is it Science?', *Discover* December: 28–52.
Vogel, G. (1997) 'Prusiner Recognized for Once-heretical Prion Theory', *Science* 278: 214.
Wright, S. (1994) *Molecular Politics: Developing American and British Regulatory Policy for Genetic Engineering, 1972–1982*. Chicago: University of Chicago Press.

Interview material

Bolton, D. (2000) Interview with author (IBR, New York: 31 July).
Bruce, M. E. (1999) Interview with author (NPU, Edinburgh: 9 June).
Carlson, G. A. (2000) Interview with author (McLaughlin Research Institute, Great Falls, Mont.: 9 August).
Carp, R. (2000) Interview with author (IBR, New York: 27 July).
DeArmond, S. (2000) Interview with author (University of California, San Francisco: 18 August).
Dickinson, A. G. (1999b) Interview with author (Dunbar: 15 September).
Outram, G. (1999) Interview with author (Edinburgh: 5 August).
Race, R. (2000) Interview with author (RML, Hamilton, MT: 14 August).
Scott, M. (2000) Interview with author (University of California, San Francisco: 18 August).

3
Paradigm Change? Explaining the Nature of the TSE Agent in Germany

Kerstin Dressel

> Discovery is not finding new things, but to look at things with new eyes.
>
> Marcel Proust

Introduction

When do knowledge claims become formal knowledge; that is, accepted as reliable, valid and useful, extending the scope of perspectives and supplying sufficient explanations for natural and/or social phenomena? What elements account for the genesis of scientific knowledge? Who are the key persons to push an idea to become a real issue on the knowledge agenda? And how can the evolutionary process be explained that a once accepted theory is superseded by a completely new way of framing an issue? These kinds of questions are addressed by scholars of the sociology of scientific knowledge (SSK), or the sociology of scientific ignorance (SSI).

One of the most far-reaching books on the evolution of scientific knowledge is Thomas Kuhn's *The Structure of Scientific Revolutions* (first published in 1962), which has been labelled by *The Times Literary Supplement* one of the 'hundred most influential books since the Second World War' (http://www.interleaves.org/~rteeter/grttls.html), and is the most cited book of the Arts and Humanities Citation Index. In this book Kuhn, professor of linguistics and philosophy at the MIT, vividly describes routines of scientific discovery and knowledge production. According to him, scientific change always follows the same course of events. A phase of normal science and its paradigm are called into question by anomalies unsolvable within the paradigm. The awareness of anomalies often leads to the application of unconventional approaches which themselves may generate new scientific discoveries. These new discoveries can then lead to a crisis of normal science and the emergence of new scientific theories. Supported by new evidence, the latter eventually bring about a scientific revolution and the establishment of a new scientific paradigm. The new, now leading, paradigm generates a new phase of normal science: a paradigm change has occurred. The cycle of scientific change is closed again.

When BSE was discovered in the UK in the mid-1980s, research into, and knowledge about, transmissible spongiform encephalopathies (TSEs) were very insubstantial, both in the UK and elsewhere. In Germany the very low incidence of TSEs[1] was the main causative element of this particular situation. One of the central and still open questions raised by TSEs is the nature of their infectious agent. Two major hypotheses are circulating around the TSE scientific community: the virus hypothesis and the prion hypothesis. Whereas the virus hypothesis has for decades been the dominant explanatory approach for the aetiology (science of the cause of diseases) of TSEs, the prion hypothesis has evolved to be the leading model of TSE research since the 1990s. None of them has been proven yet – though the prion hypothesis has been granted the Nobel Prize in the meantime.

The present chapter asks if a paradigm change has occurred in German TSE research and, if so, why. Who were the key people and what were the key assumptions of the discourse[2] that led to a paradigm change? Or, otherwise, what kinds of circumstances have prevented a paradigm change until now? This study is based on a series of in-depth interviews with German TSE researchers of both sides: supporters of the virus hypothesis and supporters of the prion hypothesis. All interviewees have studied or are still studying (at least to some degree) the aetiology of TSEs, and all of them have addressed the question of the nature of the TSE agent.

In the first section of this chapter Kuhn's theoretical framework is described. In order to contextualize German TSE discourse, in the second section this framework is applied to international developments in TSE research. Finally, a historical account of the discourse of TSE science in Germany since the late 1970s is provided with a view to determine if a paradigm change in Kuhn's sense has actually occurred.

Theoretical framework

In order to evaluate the TSE situation in Germany and to make a reasonable statement regarding the question of whether a paradigm change has taken place, we first need to focus our attention on Kuhn's theory of the 'structure of scientific revolutions'. In this section I will briefly describe the main categories proposed by Kuhn (1970 [1962]) to explain the evolution of science.

It should be stressed that my aim is not to discuss the pros and cons (Franklin, 2000; Fuller, 2000),[3] the truth, or the 'irrationalism' (Franklin, 2000) and 'inconsistencies'[4] of Kuhn's approach. Rather, this chapter is an attempt to *freely* apply Kuhn's suggestions to German TSE discourse. Differences as well as similarities between Kuhn's model and TSE discourse will be scrutinized.

Paradigm

A paradigm is a consensus shared by a scientific community, a set of recognized scientific achievements that for a certain time 'affects the structure of the group that practices the field' (Kuhn, 1970: 18). A specific paradigm is the result of diverse 'pre-paradigm schools' (Kuhn, 1970: 17) competing for the best explanatory power in a research field. From this struggle one paradigm eventually emerges to be the leading one, accepted by a scientific community as more convincing than others. It is regarded as 'more successful' (Kuhn, 1970: 23), while the other 'pre-paradigms' are left behind. The paradigm materializes in a common terminology that includes the theories explaining the paradigm, and their applications. The paradigm shared by a community defines what are the useful ways to proceed, what are the legitimate research questions to ask, which facts are relevant and which are not, what to measure and why, and how to apply which methods and procedures. Although it is the very nature of a paradigm to be accepted by the vast majority of (next generation) practitioners, single researchers may still continue to adhere to older paradigms or alternative beliefs formulated in pre-paradigms. Paradigms are closely related to 'normal science'.

Normal science

Kuhn defines 'normal science' as 'research firmly based upon one or more past scientific achievements, achievements that some particular scientific community acknowledges for a time as supplying the foundation for its further practice' (Kuhn, 1970: 10). That is, scientists working in the same field share the same paradigm and 'are committed to the same rules and standards for scientific practice' (Kuhn, 1970: 11). Normal science is any activity that goes along with the existing paradigm. It is a coherent body of knowledge shared by scientists and taught to students. In particular, normal science is, according to Kuhn, 'puzzle solving' (Kuhn, 1970: 35) and 'mopping-up operations' (Kuhn, 1970: 24). These are expressions to illustrate that neither the intention nor the goal of most researchers is to find new phenomena or new theories. As a result, innovations often go unrecognized by the scientific community because they do not fit the box of normal science paradigms. Rather, scientific efforts are made to obtain evidence that goes along with expectations of the existing paradigm, and to fill in the gaps of ignorance involved in the paradigm: '(N)ormal scientific research is directed to the articulation of those phenomena and theories that the paradigm already supplies' (Kuhn, 1970: 24).

Anomaly of normal science and the crisis of the normal science paradigm

The more puzzle solving develops the more likely that pieces will not fit in normal science, that is, anomalies are bound to appear. An anomaly is an inconsistency that seemingly clashes with the paradigm and its theories, a deviation that the paradigm is unable to answer and which, as a consequence, threatens it. Science would not deserve its name if there were no room for unconventional approaches beyond paradigms. Unconventional approaches (for instance different uses of methodologies and/or procedures) often lead to unexpected results that may constitute an anomaly within the paradigm. Alternatively, the occurrence of anomalies can initiate the application of an unconventional approach. Any scientific set-up necessarily focuses on particular aspects that appear useful and sensible in the course of normal science. Hence, other aspects regarded as outside the focus are marginalized, in particular if they incorporate conflicting evidence. Once situated in another context, however, those putatively marginal aspects and features may suddenly emerge as a pioneering result and can even lead to the formulation of a new scientific theory or even to a new paradigm. Therefore new discoveries often follow the emergence of anomalies within the existing paradigm. In any case, once the scientific community becomes aware of such anomalies, efforts are put in place either to falsify or to verify them. Verified anomalies can then actuate a crisis of the normal paradigm. The consequence of a scientific crisis triggered by an accumulation of anomalies is that scientists do not rest until they have restructured the theories of the paradigm or even the paradigm itself to ensure that former deviations are no longer unexpected but, instead, in accordance with it.

Scientific revolution

A scientific crisis brought about by the awareness of new phenomena unexplainable within the existing paradigm results in the formulation of a new theory and in changes to the basic assumptions of the paradigm. This process is called a scientific revolution. The 'transition to a new paradigm is scientific revolution' (Kuhn, 1970: 90). The framing of a new theory in order to fortify the crisis of normal science paradigms is, according to Kuhn, essential, for 'to reject one paradigm without simultaneously substituting another is to reject science itself' (Kuhn, 1970: 79).

What sounds as the normal course of events in science is in itself usually accompanied by a great deal of insecurity for researchers within the respective scientific community. 'Because it demands large-scale paradigm destruction and major shifts in the problems and techniques of normal science, the emergence of new theories is generally preceded by a period of pronounced professional insecurity' (Kuhn, 1970: 67). Scientists are no longer on the secure terrain of normal science, but instead find themselves on the shaky ground of an old paradigm that has become suspect and a new emerging paradigm that might eventually be wrong. The development of a new theory is a process that may take place over decades, sometimes even centuries (for example, the theory of the earth being a disc). The longer existing paradigms have survived without showing anomalies, the more dramatic, and, in certain cases, painful for some scientists will be the occurrence of results that do not fit the box of normal science and which cause a scientific revolution.

Kuhn is aware that scientific revolutions are more likely in some situations than in others and that they can be speeded up by external events, such as political, social, historical or technical. For instance, the astronomical crisis that Copernicus faced was accelerated by the pressure of a calendar reform (Kuhn, 1970: 69, 93). Especially important for a scientific revolution to take place is the availability of at least one (charismatic) scientist who promotes the alternative paradigm and its theories, for 'a scientific theory is declared invalid only if an alternate candidate is available to take its place' (Kuhn, 1970: 77). What Kuhn is pointing out here is the social dimension of knowledge production. In order to be successful science and its paradigms need the personality, sometimes even the charisma, of a researcher or a group of researchers. What is more, scientific revolutions are often caused by those scientists who are 'either very young or very new to the field whose paradigms they change' (Kuhn, 1970: 90).

Paradigm change

One can speak of a paradigm change when two conditions have been met. First, the new paradigm solves the puzzle more satisfactorily and explains anomalies more accurately than the previous one. Second, the new paradigm's theories are stabilized and give rise to normal science. However, Kuhn is explicit that a paradigm change is not automatically a development for good in the sense of scientific progress (Kuhn, 1970: 23). That is particularly true when considering the fact that new paradigms, regardless of being more successful than the old ones, must be interpreted only as partial explanations of a particular scientific problem. Kuhn also indicates that new paradigms are sometimes established without being preceded by a crisis.

Once a paradigm change has taken place key actors have changed, and so has the definition of the problem, its legitimacy and the way to solve it. A paradigm change leads to a new way of dealing with nature in science: '(D)uring revolutions scientists see new and different things when looking with familiar instruments in places they have looked before' (Kuhn, 1970: 111). This has a fundamental impact on scientists: '(T)hough the world does not change with a change of paradigm, the scientist afterward works in a different world' (Kuhn, 1970: 121). This is especially true of newcomers to the field who will have been trained in accordance with the new paradigm.

The paradigm change is often accompanied by the resistance of some individual scientists or groups of scientists who fiercely stick to the old paradigm because their professional life and career are grounded in it. The consequences for these researchers are evident. They will increasingly face difficulty to find funding for their work, will be completely isolated from their scientific community, and will sooner or later disappear from the field (Kuhn, 1970: 18f.).

International TSE research

There are a number of so-called transmissible spongiform encephalopathies (TSEs) in humans and animals. The most publicized and the newest is BSE, also known as 'mad cow disease'. However, sheep scrapie has the longest history.[5] It was first described in 1732 (Comber, 1772) and shown to be infectious in 1936 (Cuillé and Chelle, 1936). Creutzfeldt–Jakob disease (CJD) in humans was first described in 1920/21 by two German medical doctors and named after them. In 1976, Carleton Gajdusek received the Nobel Prize for medicine for his work on kuru, another human TSE among the Fore tribe of Papua New Guinea. Other TSEs include transmissible mink encephalopathy (TME), chronic wasting disease (CWD) in deer, Gerstmann–Sträußler–Scheinker syndrome (GSS) in humans, and fatal familial insomnia also in humans. Since the 1950s, much research has been carried out to identify the nature of the infectious agent of these diseases.

The paradigm of TSE research

Since the first third of the twentieth century it was known, from work on scrapie, that TSEs are communicable diseases. The pathogen had not been detected at that time but a consensus was shared by the TSE scientific community. The paradigm of TSE research was classical 'infectiology', that is, the knowledge that in order to establish an infection, that is, to replicate in a host, an infectious agent will use its nucleic acid (DNA or RNA). The ultimate theory of this paradigm was formulated in 1954 (Sigurdsson, 1954) when TSEs were described as 'slow virus diseases', thus indicating that the search for the infectious agent was centred on viruses with extraordinarily long incubation periods.

Normal science of TSE research

In line with the 'infectiology' paradigm normal science was looking for the virus thought to be the cause of TSE infection since the possibility of another type of infectious agent such as a bacterium, fungus or parasite, had been ruled out. Virological theories, procedures and methods were already well established and a few research groups were trying to detect the nucleic acid of a virus. Along with Gajdusek's group in the USA, researchers in Scotland also had an immense impact on the definition of TSE science since they made several important scientific discoveries on scrapie, in particular the existence of several strains of scrapie agent.[6] When the Neuropathogenesis Unit (NPU) was established in the Scottish city of Edinburgh in 1981[7] it soon became *the* research outpost for TSE research worldwide, with staff comprising some of the most eminent scrapie researchers such as Alan Dickinson, George Outram, Richard Kimberlin and later Moira Bruce – to name just a few. In line with Gajdusek's earlier work, the NPU established the viral concept as the normal TSE science paradigm. Considerable efforts were made to supply further pieces of the big 'TSE agent puzzle'. 'Mopping-up operations' in search of the virus resulted in a number of papers. Several approaches towards finding a virus were used, typical virus-specific techniques were applied, numerous scientific experiments were carried out, and repeated in modified ways, new hypotheses were formulated and eventually verified or rejected.

Although no TSE virus was identified, the virus model was nevertheless accepted by the scientific community as the leading explanation of TSEs. Normal TSE science did not consider the possibility that DNA or RNA might not be involved in these diseases.

Alas, despite all efforts, a lot of uncertainties persisted which could not be satisfactorily explained by the virus hypothesis. An expression of that situation was the reformulation of the virus hypothesis. Some scientists speculated that instead of being a virus, the agent could be a viroid or a virino.[8] However, these alternative hypotheses were not intended to break with the viral model. They emphasized the existence of known unconventional viruses (such as retroviruses and latent viruses), and did not question the presence of nucleic acid within the agent. They were firmly rooted in the 'infectiology' paradigm.

Anomalies of TSE research

The phrase 'slow virus diseases' already signalled that the TSE agent was not a conventional virus because it had a very long incubation period.[9] However, scientists were not excessively disturbed by the peculiar behaviour of this agent. But over the years a number of anomalies accumulated which threw the viral model into turmoil. We have seen that despite considerable efforts, no viral nucleic acid had been identified. Another important result that contributed to a destabilization of the viral model was the extreme resistance of the agent to heat, a rather unusual feature that cannot be explained by the ordinary virus concept.

At the end of the 1970s an important finding emerged from the unconventional approach of American neurologist Stanley Prusiner. Like other researchers in the field, Prusiner started his research by actively and systematically searching for the virus. His research involved experiments to deactivate the agent by physical, chemical, biochemical and enzymatic means. However, instead of finding the sought-for virus, Prusiner ended up with an interesting correlation. When the agent was treated with chemicals that normally deactivate nucleic acids, infectivity was maintained. Conversely, when it was treated with chemicals that digest proteins, infectivity decreased (Prusiner, 1980; Prusiner et al., 1980; McKinley and Prusiner, 1981; McKinley et al., 1983). This was indeed a pioneering and reproducible result – and an anomaly that was by no means congruent with the 'infectiology' paradigm.

Scientific revolution of TSE research

Prusiner elaborated on these results and proposed the unusual concept of a 'proteinaceous infectious particle' or, 'prion', whereby infectivity was not dependent on DNA or RNA, but caused by a protein. He published his seminal paper on the prion hypothesis in *Science* in 1982.

The reaction of the scientific community was predictable. Prusiner was initially met with amusement and strong rejection. In particular, scientists who had worked for many years (or even decades) on scrapie as a slow-virus disease were very upset.[10] This was the case of Alan Dickinson, Hugh Fraser and Richard Kimberlin at the NPU, and others worldwide as well. Their opposition was further supported by the fact that the theoretical basis of Prusiner's work was not original, leaning as it did on the earlier work of British researchers such as Tikvah Alper who, in 1967, had asked the question: 'Does the agent of scrapie replicate without nucleic acid?' (Alper et al., 1967).[11] Other scrapie researchers who had questioned the accuracy of the viral model were Gibbons (for example 1967) and Hunter (for example 1964).

Before the 1982 *Science* paper was published, Prusiner's position inside the scientific community of TSE researchers was marginal. That situation changed with this paper, for he turned the viral model upside down and developed a new theoretical framework to explain the nature of the agent. His idea of an *infectious protein* clashed with the 'infectiology' paradigm and must have appeared as a 'heretical notion'[12] to many researchers. Nevertheless, as Kuhn argues, science must remain open to unconventional approaches, at least when they appear to be able to explain what is unexplainable within the normal science paradigm. That is why Prusiner succeeded in publishing his paper in the famous journal *Science*. The TSE research community turned their efforts towards verifying or falsifying Prusiner's idea.

Prusiner's 1982 hypothesis was theoretical in nature, and lacked substantial empirical support. However, in the following years 'prion science' succeeded in delivering empirical results that stimulated the 'transition to a new paradigm' (Kuhn, 1970: 90). In 1982, Prusiner and colleagues reported the discovery of a scrapie-specific protein dubbed PrP for prion protein (Bolton et al., 1982). In 1985, a path-breaking step was made by a collaboration of Stanley Prusiner, Charles Weissmann and Leroy Hood, who discovered the prion protein gene (Oesch et al., 1985). This allowed several new experiments to be performed. Results published in *Nature* by Hsiao et al. (1989) were decisive. Hsiao and colleagues reported that a specific mutation of the prion protein gene inevitably leads to the development of GSS. In other words, GSS (and familial CJD) are caused by mutations of the prion protein gene identifiable in all carriers of the disease. This was a key finding as far as our understanding of human TSEs was concerned.

Another crucial contribution was made by Charles Weissmann's group in Switzerland. Weissmann was a distinguished molecular biologist initially sceptical of Prusiner's prion concept; however, he eventually became one of the most important prion researchers.[13] In 1992, Weissmann's group produced transgenic mice, called 'knock-out mice' (Büeler et al., 1992). These mice were genetically modified and were lacking the prion protein gene (the gene was 'knocked out'). Transmission experiments have shown that these knock-out mice do not develop disease, whereas normal mice do. Further important work was conducted by a group linked to Detlev Riesner. These researchers have clarified the size of the infectious agent. According to their results the infectious unit of TSEs, if it was a nucleic acid, would be smaller than 50 nucleotides (Kellings et al., 1992, 1995). A basic assumption of virology is that viruses are much bigger than 50 nucleotides.

Successful research and path-breaking evidence produced by the growing community of prion researchers seriously threatened the paradigm of TSE research. Importantly, much of the new evidence based on the prion hypothesis was extremely hard to reject from a viral perspective. For example, it was now possible to explain hereditary forms of TSEs (such as GSS), or to understand why the same disease could be caused in three different ways: naturally sporadic, genetically determined or acquired by contamination. These facts are extremely difficult to explain within the classical paradigm since they would require an exceptionally sophisticated and rather unusual viral model. Although final confirmation is still awaited, Prusiner's prion concept has initiated a slow, still ongoing, scientific revolution.[14] An increasing number of scientists are 'enrolled'[15] in the prion hypothesis, whereas virus supporters are now and increasingly in a minority. At the heart of the scientific revolution of the prion hypothesis lies the potential for solving previously unexplainable scientific problems.

Kuhn argues that a scientific revolution and the framing of a new paradigm are dependent upon the existence of a scientist with a strong personality, willing to fight for a new idea and able to convince others that this is indeed an important and relevant issue. He also stresses that the establishment of a new paradigm is a painful process – probably no less for the person who pushes the new idea than for those who stick to normal science. Prusiner is such a person. He was described not only as an intelligent, innovative and creative mind,[16] but was also known to be extraordinarily ambitious and obsessed by the idea of getting the Nobel Prize – almost by any means (and already as a student at the age of 23).[17] Researchers who had devoted years to studying the scrapie agent were deeply upset when Prusiner claimed that in three years he had conducted 'more experiments on the biochemistry of scrapie than everyone in the history of scrapie combined'.[18] In turn, Prusiner was severely criticized by those who follow the normal science paradigm.

Prusiner was also charged with using other scientists' findings, renaming them and claiming them to be his own. For instance, such an accusation was made in relation to the 'prion rods' discovered by Prusiner in 1983 (Prusiner et al., 1983). The same phenomenon had been observed two years earlier by Merz and colleagues and termed 'scrapie associated fibrils' (SAFs) (Merz et al., 1981). The bitterness that surrounded the scientific revolution of prions was captured by a British researcher: 'He [Prusiner] simply ignored us. It was as if work in the Thirties, Fifties and Sixties never existed. It was as if strain typing hadn't happened.'[19]

Prusiner was not the first scientist who ventured into unfamiliar territory in order to elucidate the nature of the TSE agent. However, he was the first to offer a new and complete model to a field that was longing for an explanation of the anomalies that had accumulated over the years. His heretical prion hypothesis turned out to be a fruitful approach as a number of experiments were performed and new results produced.[20] Hence, it did not bring about a punctual crisis but successfully challenged basic assumptions of the 'infectiology' paradigm. In other words, it created a scientific revolution in the TSE field.

A paradigm change in TSE research?

In Kuhn's approach, a new theory leads to a paradigm change only if it putatively solves the puzzle more satisfactorily than the established paradigm. This implies that sufficient evidence must be produced if such a change is to occur. As we have seen, Prusiner's prion hypothesis met this requirement. Many other experiments not described above due to lack of space, lend it support.

However, the one experiment that would prove beyond doubt the accuracy of the prion hypothesis has not yet been performed successfully: that is, the transformation in a test tube of the normal prion protein (PrP^{C}) into its pathogenic form (PrP^{Sc}). Another problem faced by the prion hypothesis is the existence of different strains of agent. This fact can barely be explained by the prion model, but it is comprehensible within the 'infectiology' paradigm. TSE strains and the inability to transform PrP^{C} into PrP^{Sc} are the 'thorn in the sides' of prion researchers, while they maintain a community of virus researchers. The fact that final confirmation is still awaited explains why many researchers doubt that the prion hypothesis has generated a new paradigm in TSE research.

This being said, the best indicator that the prion hypothesis is indeed threatening the paradigm of TSE research is the 1997 Nobel Prize that was awarded to Prusiner for his discovery of prions, a *new biological principle of infection*. This award is remarkable in at least two respects. Firstly, for several years the prize had not been granted to a single researcher, it was usually shared by two scientists. More importantly, the Nobel Committee had *never* awarded it to an as yet unconfirmed theory.

The peculiar situation of the TSE field makes it necessary to adapt Kuhn's approach and to introduce the notion of a 'paradigm change under construction'. On the one hand, there can be no doubt that the prion hypothesis has deeply transformed the world of TSE research. It occupies a central place in contemporary scientific literature. Students are now being trained in accordance with it and are no longer encouraged to look for a virus. Prion terminology has become hegemonic in the field. On the other hand, the lack of final confirmation is acknowledged, and the phrase 'TSE research' is still being used since it embraces both the virus and prion schools and is accepted by both sides. This paradoxical situation is captured by the phrase 'paradigm change under construction'.[21]

The TSE discourse in Germany

As we will see in the remainder of this chapter, the paradigm change under construction is also present in Germany, in a very 'ideal-typical way', to use Max Weber's phrase. CJD was discovered by Creutzfeldt and Jakob, two German medical doctors. However, for many years TSE research was underdeveloped in Germany and a proper TSE research structure did not exist until 1993. With the discovery of BSE in the UK this situation changed.[22]

The paradigm of German TSE research

Initially the paradigm of German TSE research was the same as everywhere else in the scientific community worldwide. No one doubted that infection involves nucleic acid of a pathogen and it was assumed that TSEs were caused by a virus.

In the late 1970s, two scientists in Germany were involved in TSE research. Heino Diringer, professor of biochemistry and a trained virologist, was working at the Robert-Koch Institute (RKI) in Berlin. Detlev Riesner, professor of biophysics, was at the University of Düsseldorf. Both were studying the nature of the scrapie agent. Riesner's professional networks were based mainly in the USA until 1994. In contrast,

Diringer's work was funded by German institutions or ministries (the RKI, a public sector research institute subordinate to the Federal Ministry of Health). Diringer thus appeared to be *the* German TSE expert.

Normal science of German TSE research

Diringer engaged in TSE research in 1977 on the basis of his scientific curiosity and personal interest. He was fascinated by scrapie and was intrigued by the peculiarity of the agent. Diringer collaborated with researchers at the NPU in Edinburgh. He was given research material by them and also published joint papers with them.[23] He also had good personal contacts with NPU scientists.

Diringer was a strong supporter of the virus hypothesis. He established a research group at the RKI with a view to identifying the TSE agent. The work of Diringer and his group resulted in several papers in eminent scientific journals (results of the puzzle solving of normal science). The assumptions Diringer shared with NPU researchers and others were: 'Although we don't know the disease-causing virus, it is clear that the agent is very similar to a virus in its whole conception, although it has some very special characteristics.'[24] Diringer described Merz's SAFs as an amyloidosis, that is, an aggregation of proteins in the brain caused by an 'unconventional slow virus' (Diringer, 1985). In 1983, he discovered that SAFs were composed of the prion protein (PrP) discovered by Prusiner (Diringer et al., 1983). In addition, Diringer developed a method to isolate SAFs which became a standard technique in normal TSE science. Throughout the 1980s Diringer, like the majority of researchers involved in TSE science, laid strong emphasis on the detection of nucleic acid of the putative virus.

Anomalies of research

The methods used to detect a virus are well defined and effective. Diringer was well known for conducting research in an exact and precise way. Unfortunately, he did not succeed in identifying a nucleic acid that would point to a virus, nor the virus itself. Diringer and his group had to acknowledge that the agent was in some basic respects rather unusual and certainly not 'conventional'.[25]

At the time Diringer started his research on TSE in Germany, one possible explanation for the nature of the agent discussed by the scientific community was the viroid hypothesis.[26] As we have seen above, this was an alternative to the virus hypothesis that came out of the established paradigm. Viroids are tiny causative agents, similar to viruses but lacking a protective protein coat and DNA. They are made of RNA only. Like a virus, a viroid intrudes into cells and forces the cell to duplicate its RNA instead of the host RNA. The other German TSE researcher of the time, Riesner, had just become professor of biophysical chemistry at the Technical University of Darmstadt and was interested in viroids. His first poster on viroids was published at the International Congress of Biochemistry in Hamburg and received special attention from an American neurologist also interested in viroids: Stanley Prusiner.

At another meeting, held in Germany in 1978, where Riesner was talking on sub-viral agents, Prusiner introduced his work on scrapie. By the end of the meeting, he was almost exclusively discussing the very strange behaviour of the scrapie agent. Like Prusiner, Riesner was attracted by the unusual behaviour of the scrapie agent. He and Prusiner decided to work together on the topic. In accordance with the paradigm, at that time the TSE research community mostly comprised trained virologists. Now two outsiders were working together on the scrapie agent. By 1979–80, Prusiner started to discuss with Riesner an extraordinary idea: the possibility that instead of being a virus or a viroid, the causative agent may have nothing to do with nucleic acids but be a protein. Soon afterwards, Prusiner published his first paper on the prion hypothesis in *Science* (Prusiner, 1982).

Though sceptical like most scientists, Diringer was nevertheless interested in this unconventional prion approach. His attitude changed dramatically some months later when he realized that there was something wrong with Prusiner's 1982 *Science* paper, which made the interpretation of the paper dubious to say the least. Prusiner based the prion hypothesis on the estimation that the size of the infectious scrapie agent was 50.000 M_r or less. But Diringer argued that this was an incorrect interpretation of the experiment since the concentration of the detergent used in Prusiner's high-pressure liquid chromatography (HPLC) would tend to elute infectivity[27] (Diringer and Kimberlin, 1983). As Diringer was able to show, Prusiner made a serious mistake when he exposed the sample preparation to a column that was only equilibrated with 1/400th of the detergent concentration in the sample applied. Indeed, 'the size of the infectious unit would not be a factor in elution from the column' (Diringer and Kimberlin, 1983: 565). What happened instead was, according to Diringer, the following:[28] the scrapie agent adsorbed to the column in the presence of the low concentrated detergent and then desorbed and eluted with the bulk of the detergent which happened to migrate as a discrete micell fraction with a molecular weight of about 50.000, exactly at the position where the infectivity elutes (Diringer and

Kimberlin, 1983: 564). Thus, instead of performing a molecular sieve chromatography[29] as he thought, Prusiner performed an adsorption experiment to the actual size of the infectious agent.

Diringer's serious criticism was rejected by *Science* even after an important UK expert in the scrapie field independently corresponded with the journal. Diringer then submitted a second, extended, version of his paper to *Science* which was also rejected. Finally, Diringer and Kimberlin submitted a paper to *Science* which reported an additional experiment that proved that the earlier criticism was correct. For over four months the authors did not even get an acknowledgement of receipt of their manuscript from *Science*.[30] As a result, they withdrew their paper and published it elsewhere. Their critical response to Prusiner's *Science* paper, 'Infectious scrapie agent is apparently not as small as recent claims suggest', was published in 1983 in *Bioscience Reports* (Diringer and Kimberlin, 1983). The quarrel between the prion and virus hypotheses took its course.

The following quote from an interviewee describes where most German researchers stood in relation to the prion hypothesis during the 1980s: 'In the beginning, I was not at all convinced by the prion hypothesis, like most of the other people then. Like most of the others, I thought of it as very interesting, but let's wait and see how things will develop.'[31]

Up to the end of the 1980s, German TSE discourse was dominated by Diringer, who was at the time the only TSE expert *adviser* in Germany – even if no longer the only scientist working in the field. Given that Diringer was (a) in favour of the virus concept, (b) a strong opponent of the prion hypothesis and (c) in a position of power as scientific adviser, the prion hypothesis faced much resistance in Germany. Therefore it is not surprising that the breakthrough of the prion hypothesis as a serious alternative to the existing paradigm occurred relatively late compared to the situation in the USA.

Another factor that slowed down the growth of the prion hypothesis in Germany was the chronic lack of funding at the end of the 1980s. The few highly motivated German scientists who returned from Prusiner's laboratory to continue prion research in Germany were denied research funding. An interviewee recalled the situation: 'By the end of the 1980s the opinion in Germany was more the prion theory is a "passing craze". It was almost impossible at the time to get research projects funded.'[32]

Despite initial opposition to the prion hypothesis, the paradigm of German TSE research nevertheless fell into crisis. Various German interviewees, both virus and prion researchers, stressed that in the end it was triggered by the most powerful factor: the very failure to identify the virus.

Although Diringer's critical reply to Prusiner's 1982 paper was published, Prusiner did not take any notice of it. On the contrary, he went on refining the structure of the prion hypothesis. In this endeavour he was helped by an increasing number of scientists fascinated by his intriguing new concept, including German researchers such as Riesner who started participating in *prion research* in 1984. Another German scientist (of a whole series to come) who collaborated with Prusiner[33] was the physician and later neuropathologist, Hans Kretzschmar, who was introduced to TSE research in Prusiner's laboratory in 1984. Kretzschmar attended one of Prusiner's public lectures on the prion hypothesis and was immediately intrigued by the idea that a 'prion' could be the agent of TSEs. Although he was planning to work on a different subject during his post-doctoral period, he changed his mind and joined Prusiner's group, where he also got to know Riesner. After working in Prusiner's laboratory, Kretzschmar moved to Charles Weissmann's laboratory in Zurich, with a view to continuing his research on prions. Unfortunately for him, Weissmann was still following the established paradigm and asked him to search for the virus.

When Kretzschmar returned to Germany in 1987, he was determined to pursue prion research. He had just completed his training as a neuropathologist and wanted to combine prion research with neuropathology. The biophysicist Riesner was then the only scientist working in the new prion field. The only other scientist involved in basic research on the nature of the TSE agent was Diringer who, as we have seen, opposed the prion hypothesis. Kretzschmar started to work on prions when he received a professorship in Göttingen.

In keeping with Kuhn's demonstration, it appears that German researchers who focused on the anomalies of TSE research and took up the prion hypothesis had either just received their professorships, or were new to the TSE field, and their background in biophysics or neuropathology did not immediately prepare them for work on the nature of the TSE agent.

The scientific revolution that was pushed by an external event: BSE

As soon as Kretzschmar became full professor at Göttingen University in 1992, he started to concentrate his research efforts on prions, a choice further motivated by mounting experimental evidence in support of the prion hypothesis. It had now become possible in Germany to receive funding for projects based on the prion hypothesis. The breakthrough of the prion concept and the crisis of normal science were accelerated in Germany by an external event: the BSE crisis in the UK.

Germany was confronted by the problem of how to respond to the new cattle disease in the face of scientific ignorance. As a scientist from a public sector research institute, the RKI, Diringer was already an official expert adviser of the German government when BSE emerged in the UK. His risk assessment was clear: according to the virus hypothesis, the BSE agent must be considered potentially transmissible to humans. It should therefore be regarded as a threat to public health, and should be eradicated from the human food chain. One of Diringer's papers, which reviewed publications on transmission routes of an 'unconventional virus' from one species to another, received considerable attention (Diringer, 1990). An illustration in this paper later became known as the 'Lufthansa route model', because the model of inter-species transmission was ironically compared to the route model of the German airline. Diringer described the transmission of different TSEs to other species (for instance the transmission from sheep, cattle, mink, hamster, mouse and man to other species). In showing how often the TSE agent has already jumped the species barrier, Diringer developed a scenario of BSE being a potential risk to humans and managed to convince the German federal health minister to adopt a precautionary policy to prevent the BSE agent from crossing German borders.[34]

For those researchers who support the 'infectiology' paradigm species jumping is a very likely event. But the failure to detect a pathogen based on this paradigm, combined with the generative impact and the explanatory power of the prion hypothesis, led to the final breakthrough of the latter concept on a large scale. Although Diringer was for many years (until his retirement in 1998) an expert adviser to the German government, and sat on several important EU scientific committees, his position weakened. In interviews German officials emphasized Diringer's role in stressing scientific ignorance, and also in acknowledging the possibility of BSE transmission to humans. They also stressed that neither the prion nor the virus concept satisfactorily explained BSE nor any other TSE. Nevertheless, the fact remains that at the peak of the BSE epidemic (end of 1992, beginning of 1993), most of the discussion already centred around the prion hypothesis.

By 1992–93 the scientific revolution brought about by the prion hypothesis had now reached Germany. Diringer's position in the scientific community was weakening and he was no longer the only scientist involved in advice and policy-making. In 1993, a CJD epidemiological surveillance unit was set up in Göttingen, supported by the Federal Ministry of Health and headed by Hans Kretzschmar, who enrolled Göttingen scientists into the prion concept. Around the same time the Federal Ministry of Agriculture established a national reference centre in Tübingen to monitor BSE in Germany. From the outset the work of the Tübingen centre was also carried out along the premises of the prion hypothesis.[35]

However, the best illustration that the scientific revolution was taking place in Germany lies in the 1994 creation by the Federal Ministry of Research and Technology[36] of a *TSE research programme* coordinated by Kretzschmar. The TSE research programme laid the foundation of a TSE research structure (though on a modest scale), but most importantly it stimulated *prion research activity* in Germany. Thus, after almost ten years of prion work funded by international or US grants, Riesner obtained his first grant from a *German* source for a prion research project. The vast majority of researchers involved in this programme (11 projects in 9 institutions) described themselves as prion researchers.[37]

Only 3 per cent of the budget was attributed to projects based on the virus hypothesis, that is, to Diringer and his group at the RKI in Berlin. Significantly, Diringer himself was *not* funded *to search for the virus* but to work on the transmission and pathogenesis of TSEs. In 1995, an international symposium on 'prion diseases' was organized in Göttingen by Riesner and Kretzschmar. A German interviewee described the outcome of the debate between virus and prion hypotheses as 'clear', that is, clearly in favour of the prion hypothesis.[38]

Yet, as with all scientific revolutions, opposition still existed as revealed by the following event. In the same year as the Göttingen prion symposium, the World Federation of Science invited TSE researchers to an international expert meeting held in France. In a rather informal atmosphere the question was raised of how many of the attendees were convinced by the prion hypothesis and how many were supporters of the virus hypothesis. For most participants of this meeting poll results were surprising: two-thirds of these scientists believed in the prion hypothesis, while one-third still subscribed to the old paradigm.[39]

As we have seen, Kuhn argues that 'the emergence of new theories is generally preceded by a period of pronounced professional insecurity' (Kuhn, 1970: 67). Indeed, it seems that insecurity is widespread among German researchers, including prion supporters. In an interview, a successful and eminent scientist was adamant that prion work is in any case valid, 'even if the prion theory were to prove wrong in the end'.[40]

We have seen above that until the beginning of the 1990s German TSE science was dominated by Diringer who strictly adhered to the 'infectiology' paradigm. Prion research had to wait until 1994 to receive substantial funding from national sources. However, the situation then changed rapidly. When funding from the TSE research programme came to an end in summer 2000, an extraordinary successful research programme had been completed. The programme had stimulated scientific productivity and placed German TSE research in the international arena. Most importantly, it had succeeded in bringing into Germany the scientific revolution of prions.

Again: a 'paradigm change under construction'

In interviews with German scientists the fact that Prusiner was awarded the 1997 Nobel Prize for medicine was described as 'decisive'[41] in establishing the hegemony of the prion hypothesis in Germany. The significance of this event was the same in Germany as elsewhere. Yet, scepticism has not been eliminated as shown in this comment that the viral approach 'keeps residual doubts on the prion hypothesis [still] alive'.[42] Such 'residual' doubts signal that a paradigm change is clearly under construction in Germany and this is confirmed by several factors.

First, those scientists who still subscribe to the classical paradigm do not manifest themselves. German interviewees stressed that there is probably a considerable group of scientists who do not believe in the prion hypothesis but do not express their reservations – at least publicly. As an interviewee put it: 'It is quite risky to swim against the current, because if you made your ideas explicit you won't get money for your research as it appears hopeless. You put yourself in the "camp of the stick-in-the muds", the "camp of the losers".'[43]

True, researchers at the RKI, whose work is based on the virus concept, still have a say in the field. Their work is regarded as thorough, well set up and meaningful by prion researchers who nevertheless interpret the results from their own prion theoretical framework. One factor that helps to explain the continuation of the work at the RKI is Diringer's advising role. His position in the policy world remained strong until his retirement in 1998. He received constant support from German ministries and was a member of various EU scientific advising committees on BSE, CJD and so on.

In an interview a German scientist mentioned that a modification of the viral concept towards the NPU's virino hypothesis would be 'desirable' but acknowledged in the same breath that this prospect was weak because in Germany the focus of research is no longer on the aetiology of TSEs. As he put it: 'We do have more urgent problems, like diagnosis, therapy, intervention, or prophylaxis.'[44] These topics obviously refer to the BSE epidemic and the possibility that cases of the new variant of CJD might appear in Germany too. Indeed, when the first cases of BSE were identified in Germany in autumn 2000/winter 2001, a big research consortium of more than 30 project teams was set up in Bavaria under a self-explanatory name: the Bavarian *Prion* Research Consortium (ForPrion). Unsurprisingly, the chairman of ForPrion is Hans Kretzschmar. Furthermore, on the national level, the German TSE Research Platform became established in 2001, again with Hans Kretzschmar as one of the central figures inside it.

Finally, the discourse of the German virus scientific community is increasingly open to the explanatory power of the prion hypothesis. There are three reasons for this state of affairs. First, the scientific evidence produced by prion scientists is regarded as sufficiently compelling. Second, the lack of funding for virus research is obvious.[45] Third, the new generation of TSE researchers lacks the motivation to search for the virus, not least *because* there is no funding for it. For young innovative scientists the virus approach is less attractive than the prion approach, in particular as several biotech and pharmaceutical companies have already jumped on the bandwagon of the prion hypothesis. For example, all available rapid BSE test kits from companies such as Prionics, BioRad, Enfer and InPro are based on prion research. In fact, all these companies have contacts (economic and/or personal) with Prusiner himself and his laboratory. Scientists who still follow the established paradigm are confronted with this kind of statement by their junior fellows: 'Why are you still searching for it [the virus]? Those people who worked on the prion hypothesis weren't stupid at all – they can't be wrong all together.'[46] Chamak (Chapter 5 this volume) shows that the same situation prevails in France: Prusiner is considered by young French scientists as the 'Master of the TSE field'.

Today, the TSE field in Germany is dominated by prion discourse, though eminent prion scientists admit that conclusive evidence is still lacking and that the problem of strains has not been resolved. Yet, as Kuhn observes: 'To be more successful is not, however, to be either completely successful with a single problem or notably successful with a large number' (Kuhn, 1970: 23). Although evidence clearly indicates that a paradigm change is under construction in Germany, one German scientist went one step further in an interview: 'As long as there are no new experimental hints that the virus theory might have more substance, one has to assume that a paradigm change took place. The prion hypothesis has clearly succeeded in reversing the burden of proof.'[47]

Conclusion

Although German TSE competence was initially focused on viruses, it has gradually changed towards the prion hypothesis. Despite sufficient lack of certainty, the prion concept now occupies the place of the former paradigm and has become the dominant hypothesis on the nature of the TSE agent. It is now the mainstream approach taught to students, but with a hint of the old paradigm. Interestingly, though German TSE discourse is currently characterized by a 'prion hegemony', there is still an openness detectable inside the scientific (prion) community to the possibility that a nucleic acid will be found in the end. Whereas supporters of the old paradigm stress that their work is also contributing to the prion hypothesis, even if they have a different interpretation of the results, prion researchers stress that their research will not lose validity should the prion hypothesis be falsified. However, if the majority of TSE researchers are currently working on the basis of the prion hypothesis, there appears to be a significant twilight zone of researchers who might be more in favour of the virus hypothesis, but do not want to be recognized as such. This peculiar situation of an emerging paradigm change has therefore been labelled 'paradigm change under construction'.

Whether verification of the new paradigm will be achieved one day remains an open question. At present, it seems more likely than not that the final step for the completion of a paradigm change, that is the conversion of normal PrP^{C} into the infectious scrapie form PrP^{Sc} in a test tube, will be made. Nevertheless, some international research teams are following the viroid approach. In addition, one needs to be aware that another unconventional scientific approach beyond the prion or virus concepts, conducted by a scientist perhaps not yet born, might provide the final clue to the nature of the TSE agent.

Whether or not the prion approach one day leads to a paradigm change in the full sense of the word, the assessment of the current troubled situation made by a German TSE researcher will remain: 'The terrible thing about the prion hypothesis is, which really is a revolutionary idea, once it [the prion hypothesis] is on the throne, it is dealing exactly the way the previous [virus hypothesis] did. (…) One God was deposed only to establish a new one. But probably that is a kind of necessity in the scientific process of evolution.'[48]

Acknowledgements

The idea for this chapter emerged in the context of another social scientific study I conducted for my PhD thesis in sociology at the University of Munich (under Ulrich Beck) and at Lancaster University (under Brian Wynne). The thesis was published as a book *BSE – the New Dimension of Uncertainty* (Dressel, 2002). This study was based on more than 60 interviews that I conducted with scientists, politicians, senior officials and journalists involved in the BSE case in Germany and the UK. I am extraordinarily grateful that four of the German scientists agreed to be interviewed again for this chapter (and for several subsequent telephone calls and personal communications). Without their invaluable contribution this essay could never have been written. Furthermore, all of them have read and commented on earlier versions of this chapter and have therefore safeguarded me from the worst technical failures. I would like to extend my gratitude to Jerry Ravetz, who made several important remarks on the draft version of this chapter. Moreover, I am grateful for numerous comments made by the editor, Eve Seguin, and by an external reviewer of this volume. Thanks are, again, due to Paddy van Zwanenberg for his friendly polishing of my English.

Notes

1. The incidence rate of CJD was the same as in other comparable countries: one case in a million, and scrapie was never a real problem in Germany.
2. The term 'discourse' in this chapter is used not as mere synonym of discussion, but instead in a Foucaldian meaning: a discourse embraces not only what is said, but also the practices, techniques and routines of a particular issue (Foucault, 1977).
3. A general overview on: http://www.emory.edu/EDUCATION/mfp/Kuhnsnap.html
4. For instance, Kuhn has been criticized for using the term 'paradigm' in 22 different ways. See Felt et al. (1995).
5. For a detailed background history of scrapie research, see Poulsen and Andersen, Chapter 1 this volume.
6. Scrapie has for two centuries been an issue of concern in the UK sheep flock.
7. For the history of the NPU see Kim, Chapter 2 this volume, note 1.
8. The virino hypothesis was introduced in 1979 by Outram and Dickinson – the latter would become head of the NPU in 1981.
9. Though retrospectively, one might argue that at the time knowledge about viruses was too limited regarding how rapid the replication of viruses must be.
10. See also Kim, Chapter 2 this volume, on the 'prion controversy'.
11. For more detail on the arguments raised by Tikvah Alper, see Poulsen and Andersen, Chapter 1 this volume.
12. An expression used by the journalist Emily Green in the *Independent on Sunday*, 27 April 1997.

13. Weissmann's work was so central to the development of the prion hypothesis that some scientists expected him to share the Nobel Prize for medicine won by Prusiner in 1997.
14. See Poulsen and Andersen, Chapter 1 this volume, for the factors that made the prion hypothesis recognizable and acceptable by the community. See, furthermore, Segal and Francoeur, Chapter 4 this volume, who show how graphic representations and computer models of prions helped Prusiner to be so successful and convincing.
15. An expression borrowed from the actor–network approach of Callon and Latour (1981).
16. This characterization was given by various German scientists.
17. As an anonymous German scientist recalled in an interview, Prusiner was his PhD student but left him after three weeks with the words: 'Goodbye Doc, I am going to get the Nobel Prize!' Interview with an anonymous scientist conducted in Munich, June 1999 (anonymous interviewee 1).
18. Prusiner quoted by Emily Green in the *Independent on Sunday*, 27 April 1997.
19. Quoted in the *Independent on Sunday*, 27 April 1997.
20. Eve Seguin, Chapter 6 this volume, reveals impressive figures on the productivity of Prusiner and his collaborators as expressed in scientific literature.
21. This being said, a new paradigm in the full sense of the word could soon be established. According to an anonymous German TSE researcher still engaged in the search for a virus, several virus supporters claim that they would change their mind if the results reported by Safar and Prusiner (1998) were replicated by other researchers. Telephone interview with an anonymous German scientist, conducted 19.12.2001 (anonymous interviewee 2).
22. As described by Chamak, Chapter 5 this volume, BSE has been a decisive factor for TSE research funding in France as well.
23. For instance, Diringer and Kimberlin (1983).
24. Telephone interview with Heino Diringer, 19.12.2001.
25. The term '(un-)conventional' was introduced only retrospectively.
26. Viroids were first described in 1971 by Theodor O. Diener, who discovered this pathogen to be the cause of potato spindle tube.
27. 'Elute' is a technical term which means to rinse an adsorbed substance from a column.
28. Interviews with Heino Diringer, 2.9.1998, 19.12.2001 and personal communication 11.7.2003.
29. A molecular sieve chromatography separates proteins according to their molecular size. Smaller protein molecules can penetrate the pores in the bead, whereas larger protein molecules cannot penetrate and flow down the column more quickly.
30. Heino Diringer, personal communication, 11.7.2003.
31. Interview with Hans Kretzschmar, Munich, 5.12.2001.
32. Interview with Hans Kretzschmar, Munich, 29.11.2001.
33. Interestingly, whereas German scientists tended to collaborate with Prusiner's group, or had, like Diringer, strong connections with NPU researchers, the situation in France was different: '(N)early all of the French

TSE workers visited his [Carlton Gajdusek's] laboratory and there can be no doubt that his influence on them was considerable' (Chamak, Chapter 5 this volume). In Germany, none of the 15 German researchers that were interviewed in the context of my PhD thesis, has appointed Gajdusek as a collaborator.

34. For a more detailed account of German BSE discourse, see Dressel (2002).
35. Although in an interview a German scientist described the Tübingen group as 'open' to both approaches. Telephone interview with an anonymous German scientist, conducted 19.12.2001 (anonymous interviewee 2).
36. Now the Federal Ministry of Education and Research.
37. Interviews with 15 German scientists. See Dressel (2002).
38. Telephone interview with Detlev Riesner, 11.12.2001.
39. Telephone interview with Heino Diringer, 19.12.2001.
40. Interview with Hans Kretzschmar, Munich, 29.11.2001.
41. Interview with Hans Kretzschmar, Munich, 29.11.2001.
42. Telephone interview with an anonymous German scientist, 19.12.2001 (anonymous interviewee 2).
43. Telephone interview with an anonymous German scientist, 19.12.2001 (anonymous interviewee 2).
44. Telephone interview with an anonymous German scientist, 19.12.2001 (anonymous interviewee 2).
45. Although in interviews prion researchers claimed that an innovative and well-designed virus project would definitely receive funding in Germany (for example, telephone interview with Detlev Riesner, 11.12.2001).
46. Telephone interview with an anonymous German scientist, 19.12.2001 (anonymous interviewee 2).
47. Telephone interview with an anonymous German scientist, 19.12.2001 (anonymous interviewee 2).
48. Telephone interview with an anonymous German scientist, 19.12.2001 (anonymous interviewee 2).

References

Alper, T. et al. (1967) 'Does the Agent of Scrapie Replicate without Nucleic Acid?', *Nature* 214: 764–6.

Bolton, D. C. et al. (1982) 'Identification of a Protein that Purifies with the Scrapie Prion', *Science* 218: 1309–11.

Büeler, H. et al. (1992) 'Normal Development and Behaviour of Mice Lacking the Neuronal Cell-surface PrP Protein', *Nature* 356: 577–82.

Callon, M. and B. Latour (1981) 'Unscrewing the Big Leviathan: How Actors Macro-Structure Reality and How Sociologists Help Them To Do So', in K. Knorr-Cetina and A. V. Cicourel (eds) *Advances in Social Theory and Methodology – Towards an Integration of Micro and Macro Sociologies*. Boston: Routledge & Kegan Paul: 277–303.

Comber, T. (1772) 'Real Improvements in Agriculture: on the Principle of A. Young Esq' (letters to Reade Peacock, Esq. and to Dr. Hunter, Physician in York, concerning the Rickets in sheep). London: Nicoll.

Cuillé, J. and P. L. Chelle (1936) 'La maladie dite tremblante du mouton est-elle inoculable?', *C.R. Acad. Sci.* (Paris) 203: 1552–4.
Diringer, H. (1985) 'Unconventional Slow Viruses, Amyloidosis, and Dementias', *Lancet* ii: 661–2.
Diringer, H. (1990) 'Durchbrechen von Speziesbarrieren mit unkonventionellen Viren', *Bundesgesundheitsblatt* 10/90: 435–40.
Diringer, H. and R. H. Kimberlin (1983) 'Infectious Scrapie Agent is Apparently not as Small as Recent Claims Suggest', *Bioscience Reports* 3: 563–8.
Diringer, H. et al. (1983) 'Scrapie Infectivity, Fibrils and Low Molecular Weight Protein', *Nature* 306: 476–8.
Dressel, K. (2002) *BSE – the New Dimension of Uncertainty. The Cultural Politics of Science and Decision-Making*. Berlin: Edition Sigma.
Felt, U. et al. (1995) *Wissenschaftsforschung. Eine Einführung*. Frankfurt/Main: Campus.
Foucault, M. (1977) *Sexualität und Wahrheit. Der Wille zum Wissen*. Frankfurt/Main: Suhrkamp.
Franklin, J. (2000) 'Thomas Kuhn's Irrationalism', *The New Criterion* 18 (10), June.
Fuller, S. (2000) *Thomas Kuhn: a Philosophical History for Our Times*. Chicago: University of Chicago Press.
Gibbons, R. A. et al. (1967) 'Nature of the Scrapie Agent', *Nature* 215: 1041–3.
Hsiao, K. et al. (1989) 'Linkage of a Prion Protein Missense Variant to Gerstmann–Sträußler Syndrome', *Nature* 338: 342–5.
Hunter, G. D. et al. (1964) 'Studies on the Heat Stability and Chromatographic Behaviour of the Scrapie Agent', *J. Gen. Microbiol.* 37: 251–8.
Kellings, K. et al. (1992) 'Further Analysis of Nucleic Acids in Purified Scrapie Prion Preparations by Improved Return Focusing Gel Electrophoresis', *J. Gen. Virol.* 73: 1025–9.
Kellings, K. et al. (1995) 'Analysis of Residual Nucleic Acids in Scrapie Prions with Differing Degrees of Aggregation' (dissertation). Düsseldorf: Heinrich-Heine-Universität.
Kuhn, T. (1970 [1962]) *The Structure of Scientific Revolutions*. Edited by International Encyclopedia of Unified Science 2(2), 2nd edn, Enlarged. Chicago: University of Chicago Press.
McKinley, M. P. and S. B. Prusiner (1981) 'Reversible Chemical Modification of the Scrapie Agent', *Science* 214: 1259–61.
McKinley, M. P. et al. (1983) 'A Protease-resistant Protein is a Structural Component of Scrapie Prion', *Cell* 35: 57–62.
Merz, P. A. et al. (1981) 'Abnormal Fibrils from Scrapie-infected Brain', *Acta Neuropathol.* 54: 63–74.
Oesch, B. et al. (1985) 'A Cellular Gene Encodes Scrapie PrP (27–30) Protein', *Cell* 40: 735–46.
Prusiner, S. B. (1980) 'Slow Viruses: Molecular Properties of the Agents Causing Scrapie in Mice and Hamsters', *Prog. Clin. Biol. Res.* 39: 73–89.
Prusiner, S. B. (1982) 'Novel Proteinaceous Infectious Particles Cause Scrapie', *Science* 216: 136–44.
Prusiner, S. B. et al. (1980) 'Molecular Properties, Partial Purification and Assay by Incubation Period Measurements of Hamster Scrapie Agent', *Biochemistry* 19: 4883–91.

Prusiner, S. B. et al. (1983) 'Scapie Prions Aggregate to form Amyloid-like Birefringent Rods', *Cell* 35: 349–58.
Safar, J. and S. B. Prusiner (1998) 'Molecular Studies of Prion Disease', *Prog. Brain Res.* 117: 421–34.
Sigurdsson, B. (1954) 'Rida, a Chronic Encephalitis of Sheep with General Remarks on Infections which Develop Slowly and Some of Their Special Characteristics', *British Veterinary Journal* 110: 341–54.

4
Visualizing Prions: Graphic Representations and the Biography of Prions

Jérôme Segal and Eric Francoeur

> A scientific concept that is not supported by direct visualization is always difficult to establish, whatever its origin may be.
>
> D. Dormont

Prions are proteins generally characterized by the ability to exist in two different forms or more precisely two different three-dimensional structures, one of them possibly causing disease when it aggregates. The prion hypothesis, as formulated by Stanley Prusiner, states that this aggregation causes specific neurological diseases such as bovine spongiform encephalopathy (BSE). Even if both the mechanisms of this change of conformation and that of the aggregation are still enigmatic, the prion hypothesis has become a dominant model to which much heuristic power has been attributed in the 1990s. This could be a first paradox.

Moreover, whereas three-dimensional structures clearly appear to be at the heart of the matter, Prusiner used mostly biochemical evidence to develop his hypothesis, without using, in the early days, any other graphic representations than that given by electron microscopy. This constitutes the second paradox at the origin of the present chapter since only computer representations of three-dimensional structures can explain and justify the prion theory as a model. Here, models are defined as theories with two distinct properties. First, models have an explanatory power more or less confirmed by experimental evidence, which distinguishes them from mere hypotheses. Second, models can be applied in domains other than those where they come from. Such application is possible due to the underlying formalism of models, or, as in the prion case, to the diffusion of a specific visualization culture.

Since the second half of the 1990s, many scientific journals have published three-dimensional representations of prion structures, always in the non-pathological form (the structure of the other form remaining as yet unknown). What are these representations supposed to bring? Why were they published, sometimes on the cover of prestigious journals? How have they been obtained? The use of computers has of course been decisive but more generally, did the visualization aids provided by bioinformatics help to change the epistemological status of the prion hypothesis?

To tackle these questions, we will first outline the context in which three-dimensional structures of proteins, historically called 'tertiary structures', have become an important scientific topic. We will then review the place of graphic representations in Prusiner's work, and show how the prion hypothesis has changed metaphors in biology. This will allow us to concentrate on the conformational change, and to end our narrative with an analysis of the main ongoing projects on the tertiary structure, with particular emphasis on the case of 'yeast prions'.

An overview of molecular visualization

Until fairly recently, historians and philosophers of science paid scant attention to the issue of visual representation in science.[1] Since the mid-1980s, scholars in science studies have become increasingly concerned with the role of visualization and visual representation in the development and practice of science.[2] From this literature has come the clear conclusion that visual representation is far from an epiphenomenon of scientific practice, but rather one of its intrinsic elements.

The issue of visual representation can be understood not only in terms of techniques and technology but also in terms of the various practices and activities associated with making 'natural' objects observable and intelligible. Ethnographic studies of laboratory activities have been particularly instructive in this regard, showing the transformation over time of research objects and their gradual shaping into pictorial data and graphic displays.[3] The present chapter deals with visualization in molecular biology.

The recent completion of the Human Genome Project in 2001 has brought disappointment for all those who believed it would lead to rapid progress in gene therapy or at least provide a better understanding of protein synthesis. This worldwide project has produced the complete sequence of human DNA. As is well known, this DNA sequence codes for the amino acids that constitute proteins. The 'primary structure' of proteins is given by this sequence of amino acids, and the assembly of some regular structure, such as alpha helices and beta sheets, defines the 'secondary structure'.[4] If these structures can be defined without complex graphic representations, the full 'tertiary structure' of proteins, their functional three-dimensional shape, cannot be easily described in its full complexity without visualization devices since the secondary structure only gives hints to the arrangement of the tertiary structure (only parts that are identified as helices and so on). The problem of 'protein folding' corresponds to the process during which the protein acquires its tertiary structure.

The emphasis laid on DNA in the 'Central Dogma'

In 1957, four years after publishing with James Watson the structure of DNA, Francis Crick held a conference 'on protein synthesis' (Crick, 1958).[5] He clearly stated why he chose to concentrate on the primary structure:

> Our basic handicap at the moment is that we have no easy and precise technique with which to study how proteins are folded, whereas we can at least make some experimental approach to amino acid sequences. For this reason, if for no other, I shall ignore folding in what follows and concentrate on the determination of sequences. (Crick, 1958: 144)

Even if he insisted at the beginning of his talk on the fact that, as in the case of enzymes, proteins owe their specificity and activity to the properties of their tertiary structure, Crick dealt mostly with 'information', which he defined as 'determination of sequence, either of bases in the nucleic acid or of amino acid residues in the protein'. He developed his views under the hypothesis that 'folding is simply a function of the order of the amino acids' and the conference became famous because of the formulation of what he called the 'Central Dogma' of molecular biology. As he wrote, 'This states that once "information" has passed into proteins it cannot get out again', which means that information cannot flow from proteins to genes (Crick is somewhat vague about the role of RNA which determines the sequence of amino acids).

Crick later explained that he meant 'axiom' rather than 'dogma' but the diffusion of this idea led many researchers to consider the analysis of DNA as a quest for the Holy Grail. Under the assumption that DNA sequences would explain protein synthesis, most funding went to genetics and protein studies became somewhat neglected. A static conception of protein dominated, whereas biochemists knew that the study of the folding process requires a dynamic approach.

Interactive molecular graphics and the heuristic role of computers

Molecular biology has been developed in a civilization based on writing. Scientists publish articles and even when they talk, they say that they present 'papers'. How did they acquire and transmit their knowledge regarding protein tertiary structures? To understand the structure of molecules, biologists managed – often with the aid of X-ray analysis – to build physical models of the molecule they wanted to study. Robert Corey and Linus Pauling, who offered tools to identify secondary structures, designed various types of models. Some of these models emphasized the volumes occupied by atoms in the molecules and afforded an understanding of steric hindrance. The Corey–Pauling–Koltun space-filling models, based on an original design by Corey and Pauling, became very popular in the late 1960s (Francoeur, 2001).

These physical models did not allow for satisfactory manipulation and their construction often proved physically impossible for big molecules (some biologists even contemplated building models under water to avoid the effect of gravity). The breakthrough came with the development of time-shared mainframe computers, which allowed real-time functioning and interactivity between the user and the machine. The precise origin of the concept and techniques of interactive molecular graphics can be traced back to a group of scientists around the molecular biologist Cyrus Levinthal (1922–90), active at the MIT in the mid-1960s (Francoeur and Segal, 2004).

Interactivity was the key element of his visualization device called the 'Kluge'. It referred to the relative ease with which the scientist was able to transform the display to highlight particular features of the displayed object or modify specific parameters of a simulation and get a fast or immediate visual feedback. In short, this interactivity implied a capacity to experiment and tinker with the data being modelled or the phenomenon being simulated. Skilled scientists learned to see what was being disclosed and in this sense, interactive molecular graphics became a way of revealing the inner character or hidden nature of things.

Because the Kluge was a vector-based display, only the bonds between the atoms could be represented, recreating the visual experience of skeletal models, without the problems of gravity. The illusion of three-dimensionality was created by rotating the structure on the screen and having the user control the rate of rotation through the 'track-ball'. A light-pen and buttons were also used to interact with the displayed structure.[6]

In the 1960s, the use of computers by Levinthal's team did not lead to important scientific success. For example, the structure of cytochrome-c he proposed was not the same as the one obtained with classical methods of crystallography.[7] However, in a long-range historical analysis, it appears that this work deeply changed the work of biologists interested in the tertiary structure of proteins. At this point, biochemistry met computer science and some of Levinthal's collaborators like Martin Zwick remember that they were quite happy to leave 'wet chemistry in favor of computation'. On a sociological level, a new community was born around objects that were no longer real, but digital representations of molecular structures. Most of Levinthal's co-workers helped promote interactive molecular graphics and alongside the expressions *in vivo* and *in vitro* which already characterized biological studies, a new approach was introduced: *in silico*.

Many different programmes and visualization devices were developed between the 1960s and the early 1980s when Prusiner introduced the prion hypothesis. In the mid-1970s for instance, a first protein structure was solved by means of crystallography and visualized entirely with computers (without building a physical model).[8]

The scope of this chapter does not allow us to comment on the place of visualization in all the different works on protein structure. The important characteristic of the history of molecular visualization is that a co-evolution exists between the state of the knowledge and the representation of structures. For example, when the relevance of describing secondary structures with alpha helices and beta sheets was admitted, schematic conventions to represent these structural elements were adopted. In this sense, we will try to show in the following sections how representation determines current knowledge related to prions, keeping in mind that specific modes of visualization 'frame' the thinking about the object they represent. Time has now come to look at how prions have been represented, bearing in mind that representations are a product of scientific activity and also influence the way science is being done. Our aim is to see how these representations affect prions as epistemic things, which at the same time result from investigations and steer their course, until they finally settle into well-defined concepts.[9]

Representations of Prusiner's prion hypothesis

The function of representations in Prusiner's work in the 1980s

In his 1982 publication in *Science*, Stanley Prusiner introduced the word 'prion' to denote 'small proteinaceous infectious particles'. The methods he used belonged to a large extent to biochemistry and also to virology for the study of infectious properties (Prusiner, 1982).[10] The major question concerned the way in which prions 'replicate', if they are devoid of nucleic acids. An 'interesting analogy' was made with retroviruses (where in a schematic way, information flows from RNA to DNA), and also with the 'auto catalytic' property of the tobacco mosaic virus. For the most part, Prusiner's theory was based on the long observation of diseases like scrapie, kuru and Creutzfeldt–Jakob disease (CJD). Determining the molecular structure of prions was only considered as a means to gain better understanding of the aetiology of these diseases: 'A knowledge of the molecular structure of prions may help identify the aetiology of some chronic degenerative diseases of humans' (Prusiner, 1982: 143). At that time, the idea that a molecule could exist in two conformations, one of them being able to aggregate, was not mentioned.

In a review article Prusiner published in 1984, the main issue was still the replication or reproduction of prions in the absence of nucleic acids (Prusiner, 1984).[11] The question asked in relation to this 'biological conundrum' was nothing more than 'what is the nature of their genome?' (Prusiner, 1984: 48). Prusiner had tried to isolate the infectious agent and produced pictures. On the second page of the paper, we find two micrographs (Figure 4.1) with the following caption:

> *Prions in the brain* of a hamster are identified by an immunological staining technique. The hamster had been infected with scrapie, the prototypical prion disease, which in nature affects sheep and goats. After an incubation period of roughly two months a section of brain tissue was exposed to antibodies with a specific affinity for a protein called PrP, the major constituent of the prion, and possibly the only constituent. (Emphasis in original)

This text is worth quoting entirely since it raises the question of the nature of the observable: it is not prions that are directly observed but only antibodies, which have the specific property of reacting with prions. The antibodies used in the preparation were labelled with an enzyme (peroxidase) which, as Prusiner explained, 'catalyses the conversion of a colourless reagent into a dark stain' (Prusiner, 1984: 49). The micrographs showed these stained structures which were not actually 'prions'. Electronic microscopy was used to show what was thought to be 'aggregations of prion "rods"', described as 'tufts with a fluffy texture'. These rods were supposed to be 'a condensation of perhaps 1,000 PrP molecules' and the fact that they were indirectly represented (with specific antibodies) helped to stabilize the theoretical existence of prions as infectious agents devoid of nucleic acids. Thanks to these micrographs,

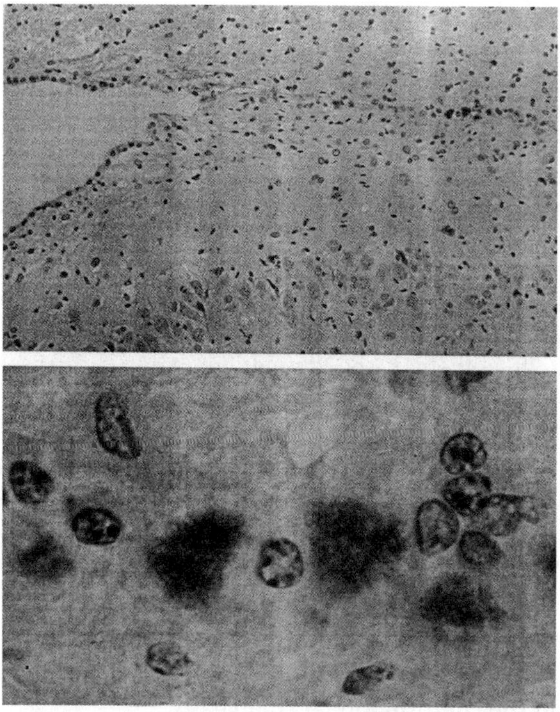

Figure 4.1 Micrographs used by the Prusiner group showing prions that reacted with antibodies

Source: Prusiner (1984). (Reprinted with permission from Stanley Prusiner and from *Scientific American* 251: 49. © 1984 *Scientific American*.)

the main issue could shift from the search for nucleic acids to the structural study of these rods, which were noticed to 'closely resemble amyloid plaques', specific to diseases such as Alzheimer's. A scientific culture related to visualization devices then emerged not only in Prusiner's prion research but more generally in the TSE field.[12]

As shown by Seguin (Chapter 6 this volume), many of Prusiner's papers were published in journals traditionally devoted to neurodegenerative diseases: *Neurology, Annals of Neurology, Journal of Neuropathology and Experimental Neurology*, or *Acta Neuropathologica*. It was in this latter journal that Prusiner published in 1987 a review article in which he specified the infectious part of PrP (PrP27-30), and concluded that 'the conformational differences between PrP^C [the "cellular" protein, non-pathological] and PrP^{Sc} [the protein that causes sheep *scrapie*] are unknown but probably arise from post-translational modifications' (Prusiner et al., 1987: 299). In this paper, prion rods were shown on three different scales. These illustrations were much more detailed, and their composition was also given: rods are aggregates of PrP27-30.

A step further was then taken in the structural analysis of PrP. In a paper published in 1988, Prusiner indicated that PrP was made of 254 amino acids, and concluded that 'defining the chemical and/or conformational differences between PrP^{C} and PrP^{Sc} is of paramount importance, as is learning how to synthesize biologically active prions' (Prusiner, 1988: 117). At that stage, the existence of prion nucleic acid was qualified as 'hypothetical' and Prusiner proposed a diagram (Figure 4.2) to visualize three main hypotheses of the conformational change: (a) the existence of prion nucleic acid, (b) the modification by PrP^{Sc} of the gene encoding PrP, or (c) the self-triggering of PrP^{Sc} to induce a conformational change in PrP^{C}. The caption indicated that this diagram illustrates 'three possible models of prion multiplication'.[13] The word 'multiplication' instead of 'reproduction' or 'replication' previously used, clearly showed that the first hypothesis was given less and less credit.

Trying to predict the secondary and tertiary structures to understand the conformational change

The secondary structure of PrP was proposed by Prusiner and his team in 1992, based on biochemical models (Gasset et al., 1992). Prusiner and colleagues used synthetic peptides reproducing the four parts of PrP in which they hypothesized the existence of α-helical regions. Three out of the four synthetic peptides formed amyloid plaques composed largely of β-sheets. Hence, Prusiner and co-workers came to the idea that the putative conformational change between PrP^{C} and PrP^{Sc} was due to a change of α-helices into β-sheets.

Micrographs were shown to illustrate the authors' hypothesis on the secondary structure (Figure 4.3). The regions that 'might form α-helices under monomeric conditions' were designated H1, H2, H3 and H4. The caption read: 'Electron micrographs of H1(...), H3, H4'. In fact, this caption was rather misleading since only aggregations of polymerized peptides were displayed.

One may wonder about the relevance of reproducing these micrographs. Were they regarded as visual proof that PrP^{C} contained three or four helices that could change into β-sheets in PrP^{Sc}? Since they did not demonstrate it, one could argue that they actually weakened this theory.

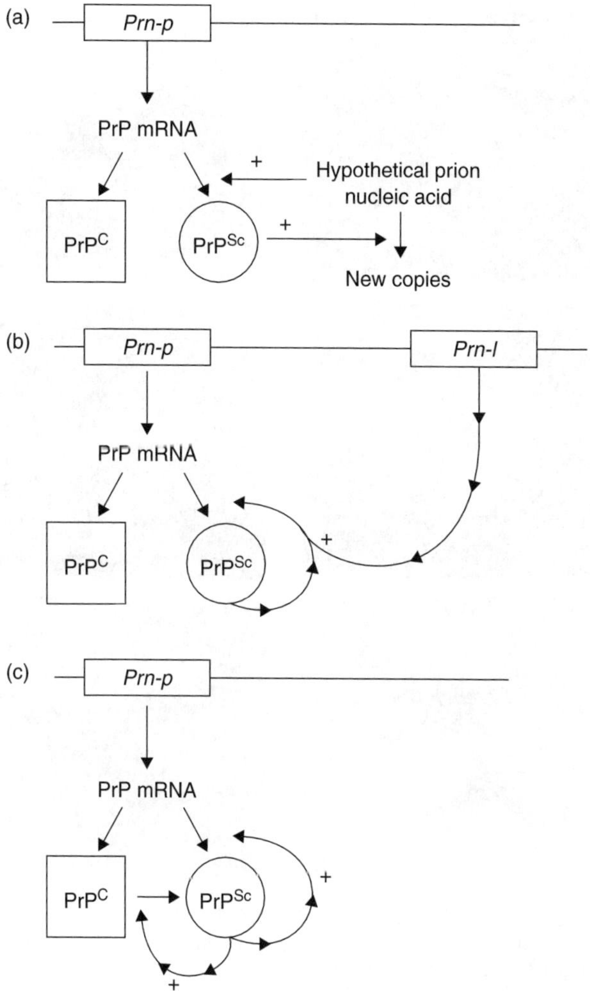

Figure 4.2 Three possible models of prion multiplication

Source: Prusiner (1988). (Reprinted with permission from *Advances in Virus Research* 35: 121. © 1988 Elsevier.)

In any case, the prediction of the secondary structure of PrP^C had two important consequences. On the one hand, it allowed Prusiner and colleagues to better characterize the change of conformation. In the same year, the diagram used to illustrate the multiplication of PrP^{Sc} (Figure 4.4) was much more univocal compared with the previous one (Figure 4.2)

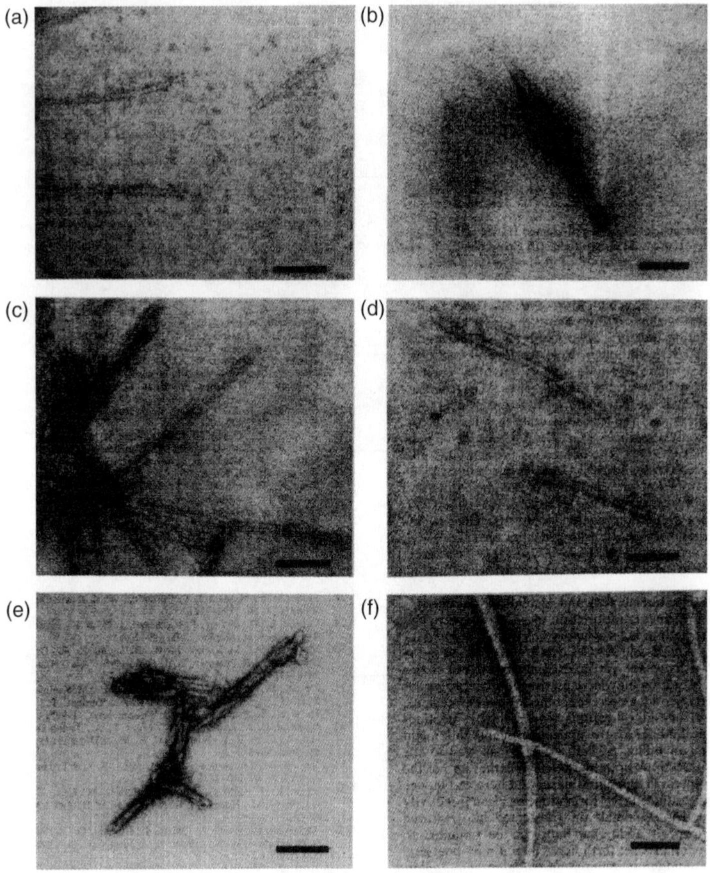

Figure 4.3 Electron micrographs of polymerized peptides

Source: Gasset et al. (1992). (Reprinted with permission from *PNAS USA* 89(22): 10943. © 1992 National Academy of Sciences, USA.)

On the other hand, Prusiner's prediction of the secondary structure of PrP^C shifted the debate away from the notion that PrP^C could replicate without nucleic acid. Scientists then started to focus on this prediction, and some disputed it.[14]

In 1992, a specialist in computational biology, Fred E. Cohen, had joined Prusiner's team. In 1994, two years after the publication of the predicted α-helical regions, they proposed a three-dimensional structure of PrP^C (Huang et al., 1994). In this paper, the PrP used was common to

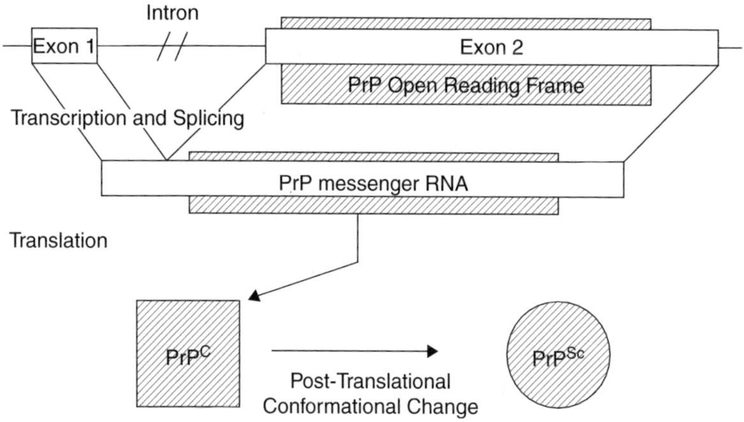

Figure 4.4 Multiplication of PrP^{Sc}

Source: Prusiner (1992). (Reprinted with permission from *Biochemistry* 31(49): 12278. © 1992 American Chemical Society.)

different species. We read that 'PrP amino acid sequences from 1 avian and 11 mammalian sources including chicken, cow, sheep, rat, mouse, hamster, mink and human were used' (Huang et al., 1994: 7139).

Prusiner and colleagues' prediction of the three-dimensional structure of this PrP^{C} was presented as a result of computational studies and for the first time 'computer modelling' featured in the keywords. Following up the 1992 paper, the researchers tried to explain the stable tertiary structure of PrP^{C}. The transition from the secondary to the tertiary structure was made by 'exploiting recent advances in protein structure prediction algorithms' in order to obtain a three-dimensional structure of PrP^{C} 'based on a family of homologous amino acid sequences' (Huang et al., 1994: 7139).

This notion of homology depended on the constitution of databases like ExPASy (*Ex*pert *P*rotein *A*nalysis *Sy*stem) at the Swiss Institute of Bioinformatics whose Internet server had just been opened (1 August 1993).[15] When the 1994 paper was published structure predictions were being tested and compared to crystallographic and nuclear magnetic resonance (NMR) studies.[16] In December 1994 the first meeting on Critical *A*ssessment of *t*echniques for protein *S*tructure *P*rediction (CASP) was organized in Asilomar. The idea was to organize a contest between computer models, differentiating three topics: 'comparative modelling', 'fold recognition or threading', and '*ab initio* folding'. Fred Cohen,

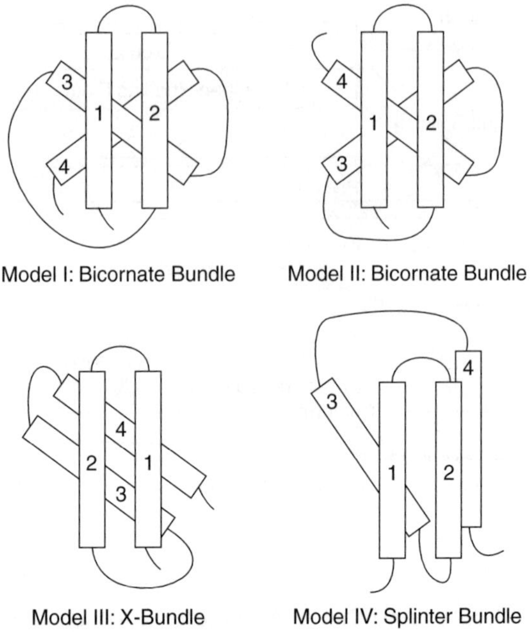

Figure 4.5 Four kinds of helices for possible structures of PrP^C

Source: Huang et al. (1994). (Reprinted with permission from *PNAS USA* 91(15): 7141. © 1994 National Academy of Sciences, USA.)

co-author of the 1994 paper, was in charge of the last topic (Huang et al., 1994).[17]

In the 1994 paper, the four helices identified two years earlier were arranged in a three-dimensional structure (Figure 4.5). Among four possible topological arrangements an X-bundle structure was chosen based on minimal distances between helices, even though the authors explained that the algorithms they had used to predict the secondary structure were probably not appropriate for a protein that exists in two conformational isoforms.

Based on the X-bundle structure, a three-dimensional structure of PrP^C was proposed by Prusiner and colleagues (Figure 4.6). In Figure 4.6, we see the same structure twice but with different indications. On the left-hand side, the predicted helice interaction sites are highlighted, whereas on the right-hand side, we see the mutation points that were supposed to explain genetically the difference between PrP^C and PrP^{Sc}. This figure was used as a means to understand how the conformational

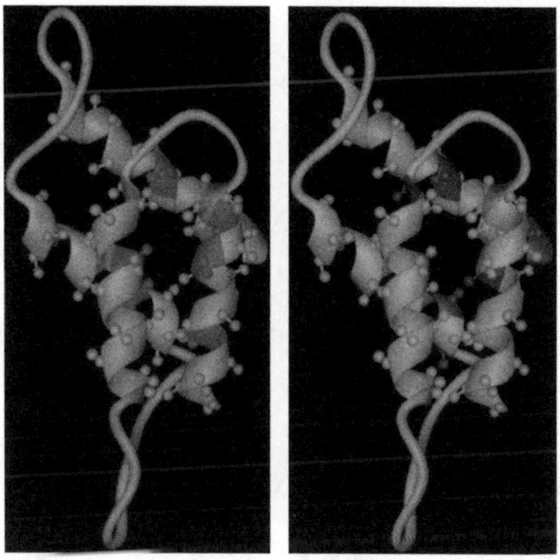

Figure 4.6 Stereoview of the predicted three-dimensional structure of PrP^C (the points of helix–helix interactions are highlighted). The original colour figure can be found at http://prions.free.fr

Source: Huang et al. (1994). (Reprinted with permission from *PNAS USA* 91(15): 7142. © 1994 National Academy of Sciences, USA.)

change could take place. The caption stated: 'we suggest that these mutations destabilize the structure of PrP^C and thereby facilitate the conformational change that features in the formation of PrP^{Sc}' (Huang et al., 1994: 7142).

In a review article also published in 1994, the prion was no longer presented as a hypothesis but rather as a 'concept' (Prusiner, 1994). The section 'development of the prion concept' was placed just before the section on the discovery of the prion protein and Prusiner was proud to announce that 'after a decade of severe criticism and serious doubt, the prion concept is now enjoying considerable acceptance' (Prusiner, 1994: 658).

From the mid-1990s onwards, graphic representations of the three-dimensional structure played an increasing role in prion research. Differences in the three-dimensional structures of PrP^C and PrP^{Sc} became centre stage. Just as he had proposed a structure of PrP^C in 1994, Prusiner proposed a three-dimensional structure of PrP^{Sc} in 1996 (Huang et al., 1996). Prusiner and colleagues first chose among six topological arrangements and then used databases. Two figures in this 1996 paper used the same conventions as those used two years earlier (Figures 4.7 and 4.8).

Visual representations of PrP^C and PrP^{Sc} led to a better understanding of the conformational change, linking the results obtained by genetics

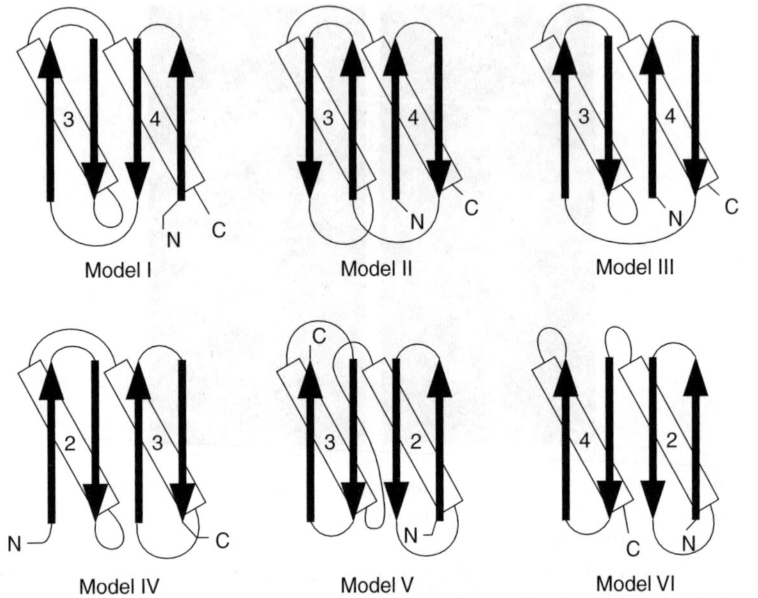

Figure 4.7 Schematic drawings of the six plausible structures of PrP^{Sc}
Source: Huang et al. (1996). (Reprinted with permission from Stanley Prusiner and from *Curr. Top. Microbiol. Immunol.* 207: 57–8. © 1996 Springer.)

and biochemistry to those obtained by computer modelling. The outcome was shown by superimposing two diagrams, one representing a classical chemical process and the other representing three-dimensional structures (Figure 4.9). Figure 4.9 introduced spatiality to gain a better understanding of prion structure and, compared to Figures 4.2 and 4.4 above, marked a transition in the epistemological function of representations.

To try to explain the conformational change and show more precisely the possible interactions between parts of the three-dimensional structure of PrP^{C}, namely helices, Prusiner decided to reinforce the biocomputational part of his research. In a paper published in September 1997, he proposed the 'Solution structure of a 142-residue recombinant prion protein' of a Syrian Hamster (SHa) (James et al., 1997). Using NMR, Prusiner took advantage of visualization software developed in Switzerland by Kurt Wüthrich and his team (program Diana) to represent the data (Güntert et al., 1991).[18] In this paper, four of the five representations were made in stereoviews (Figure 4.10). A stereoview is

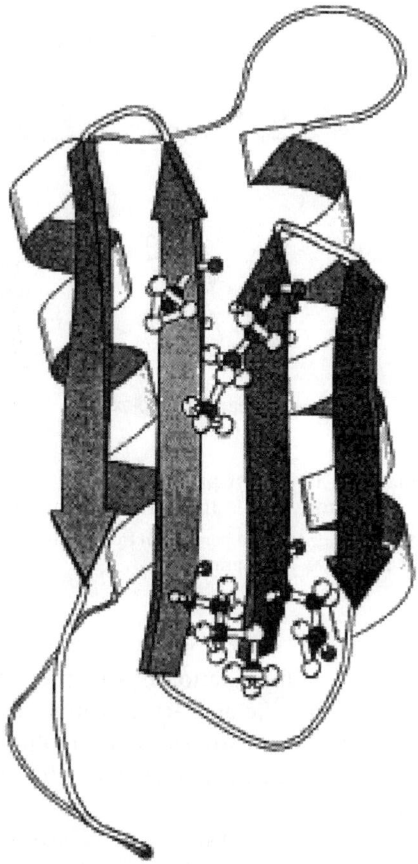

Figure 4.8 Proposed three-dimensional structure of PrP^{Sc} after correlation with genetic data on residues involved in species barrier

Source: Huang et al. (1996). (Reprinted with permission from Stanley Prusiner and from *Curr. Top. Microbiol. Immunol.* 207: 58. © 1996 Springer.)

made up of two almost identical photographs mounted side by side on a card. Looking at a stereoview through a special viewer gives the impression of a single, three-dimensional image.[19]

As the original caption of the stereoviews indicated, these graphic representations displayed information that could not be conveyed in textual format.[20] Such tangling up of alpha helices and beta sheets was too complex to be described and scientists had to be trained to decipher stereoviews. The analysis of the proposed three-dimensional structure

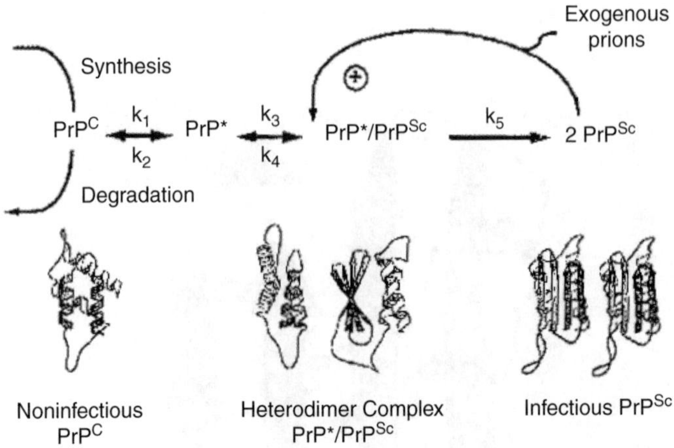

Figure 4.9 Conformational model for prion replication (PrP* is a rare partially unfolded, monomeric structure, that is an intermediate in the formation of PrP^{C})

Source: Huang et al. (1996). (Reprinted with permission from Stanley Prusiner and from *Curr. Top. Microbiol. Immunol.* 207: 61. © 1996 Springer.)

(Parts 'D' and 'E' of Figure 4.10) correlated with a study of PrP^{C} in different species led to the formulation of the 'protein X' hypothesis, which would be species-specific and act as a molecular chaperone in PrP^{Sc} formation.[21] This demonstrates that graphic representations can play an active role in scientific work since they can lead to new hypotheses and models.

In October 1997, Prusiner was awarded the Nobel Prize for physiology or medicine. He was awarded it on his own (which had not happened for ten years) and his theory was not yet proven. This gave rise to criticisms, as shown by Kim (Chapter 2 this volume). The fact remains that on the sociological level the effect of the award was to reinforce the validity of the prion concept. In his Nobel lecture, Prusiner used the pictures of a modelled three-dimensional structure displayed above and these images benefited from widespread exposure (Prusiner, 1998).[22]

Soon afterwards, further emphasis was laid on graphic representations. Until then, protein studies had been dominated by a rather static approach: in line with Anfinsen's theories, the folding process was characterized by its initial and final states (Anfinsen, 1973).[23] In contrast, the existence of a conformational change was now leading to a new dynamical approach. In terms of visualization, this new approach made the presentation of results in printed format difficult. Colleagues interested

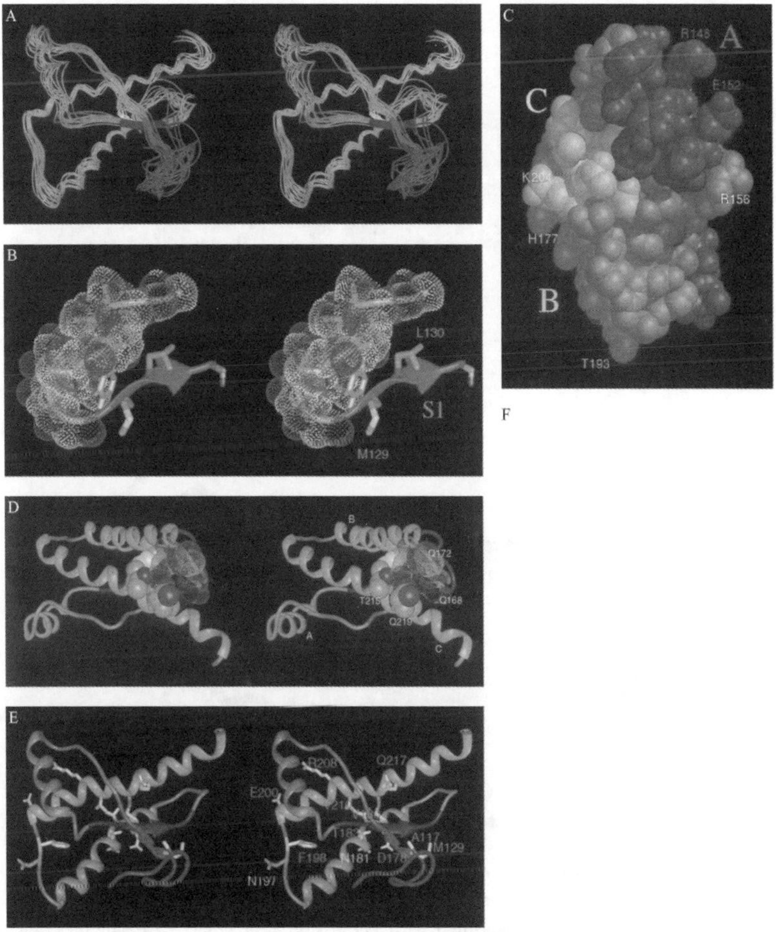

Figure 4.10 Stereoviews of an NMR structure of a Syrian Hamster prion (the different views show different parts and, in the original publication, colours are used to highlight different secondary structures). The original colour figure can be found at http://prions.free.fr

Source: James et al. (1997). (Reprinted with permission from *PNAS USA* 97(19): 10088. © 1997 National Academy of Sciences, USA.)

in this kind of work had to experience the proposed conformational change on a screen. A paper published in December 1997 reported on the flexibility of a recombinant PrP. In the abstract, Prusiner and colleagues announced that 'detailed information about PrP^C structure may provide essential insights into the mechanism by which these diseases

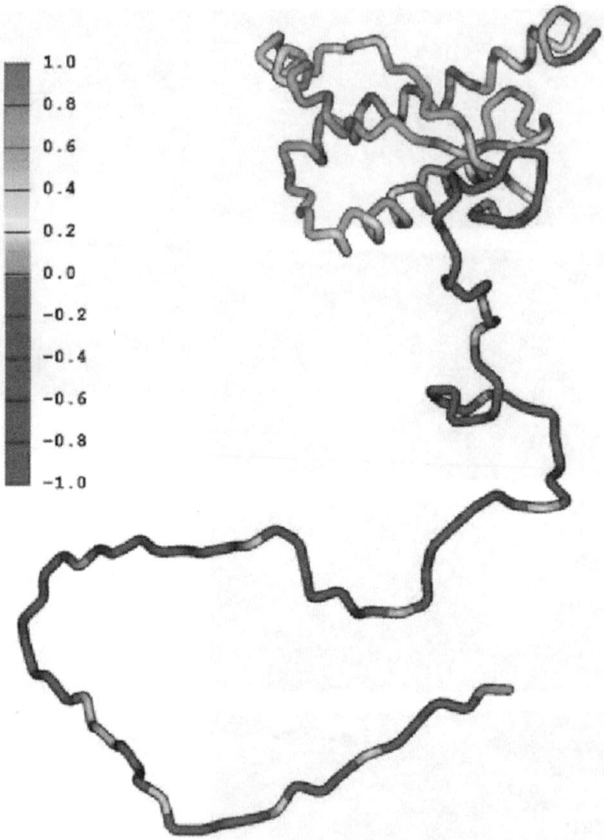

Figure 4.11 Schematic diagram showing the flexibility of the polypeptide chain for PrP (29-231). The original colour figure can be found at http://prions.free.fr

Source: Donne et al. (1997). (Reprinted with permission from *PNAS USA* 94(25): 13456. © 1997 National Academy of Sciences, USA.)

develop' (Donne et al., 1997: 13452). The announced 'detailed information' was provided in a figure where the flexibility levels corresponded to a colour scale (Figure 4.11).

The three-dimensional structure of PrP^C was used to convey information aimed at explaining the conformational change in its *dynamical* aspects. It is usually difficult to represent in one's mind the flexibility of proteins. In the last five decades, structures have been extensively represented, whereas dynamical properties were neglected. Though in the early 1960s Levinthal and colleagues produced films, these were not widely diffused. As a result, when seeing a protein researchers used to think only in terms of structure. In contrast, thanks to the use of different colours, Figure 4.11 represents the ability to move, that is, readers can imagine the movements of the different parts of PrP^C. In this sense, from the adoption of the dynamic approach, diagrams shaped the reader's mind. This exemplifies how representations of prions have become epistemic things: diagrams are presented as the outcome of research but they also influence the way researchers define their object.

This brief review has shown that in order to develop and foster the prion hypothesis, Prusiner increasingly resorted to a range of visual devices, from micrographs of prion rods to computer modelling of PrP. Colourful graphic representations generated by computer have featured on the cover of a number of journals, and in this way the prion hypothesis became so popular that it could serve as a model. In particular, computer modelling enabled precise representations that permitted a better insight into the conformational change. In the following section we will see that the progressive diffusion of this kind of work has changed the way biologists interested in protein studies and TSEs use metaphors.

The use of metaphors

Metaphors have different statuses in science. If some of them are recognized as such, others derive from the development of predominant scientific discourses and are used somewhat unconsciously.[24] The development of the prion hypothesis and the popularization of related representations provide a unique opportunity to analyse a shift in the use of metaphors in molecular biology. The 'informational metaphor' has been progressively replaced by other, more concrete, metaphors like the 'domino-stone'.

The power of the informational metaphor

Traditionally the explanation of the process of infection has been dominated by a discourse based on information theory. Information theory derived from the mathematical theory of communication and from cybernetics, and had applications in many different fields. This development was concomitant with the discovery of the DNA structure. In the lapse of time between the publication of Crick and Watson's paper (1953) on the double helix and the identification of the genetic code (1961), information theory deeply influenced discourse production in molecular biology. Kay (2000) has shown that even though on the scientific level information theory has been of little help, it has nevertheless generated important informational metaphors. Genetic 'information' was at the core of a number of studies. The discovery of retroviruses in the early 1970s (for which Howard Temin received the 1975 Nobel Prize) did not really change this conceptualization since scientists were still speaking of an information flow, though from RNA to DNA.

In academic journals devoted to the history of science, only one article has been published on prion history, which explores the challenge of the 'Central Dogma' of molecular biology by the prion hypothesis (Keyes, 1999). However interesting it may be, the discussion is rooted in a misunderstanding of the notion of information in biology. Keyes seems unaware of the metaphoric nature of the notion of information. Thus, in addition to the classical 'sequential information', she proposes the concept of 'conformational information': 'a possible new method of replication achieved via the transfer of *conformational* information forced a reassessment of the elements of molecular biology's theoretical framework' (Keyes, 1999: 4). Unsurprisingly, she also defines the prion as an 'information molecule' and grants it an 'informational role' (Keyes, 1999: 210).

In stark contrast, other authors, mostly scientists, found in the prion theory an opportunity to get rid of this information metaphor and chose to illustrate their point of view with graphic metaphors that differ from the traditional arrows illustrating information flows.

The 'domino-stone' metaphor

As it became clear that thinking in terms of genetic information was not relevant to the understanding of prion diseases, other metaphors were developed to explain the spread of PrP^{Sc}. In 1996, Adriano Aguzzi at the Institute of Neuropathology (Zurich University Hospital), and Charles Weissmann at the Institute of Molecular Biology (University of Zurich), worked on PrP^{C} and showed that this molecule was required for infection by PrP^{Sc}. Studying the 'propagation of the infectious agent', they gave up the information metaphor and introduced the 'domino-stone' metaphor: 'Within the framework of the protein-only hypothesis, these findings [the fact that PrP^{C} is required for the spread of scrapie] may be accommodated by a "domino-stone" model in which spreading of scrapie prions in the CNS [central nervous system] occurs *per continuitatem* through conversion of PrP^{C} by adjacent PrP^{Sc}' (Brandner et al., 1996: 13151). This shift was so important to them that they made a 'model' out of it, which has been extensively used by Aguzzi's team to study the conversion of PrP^{C} into PrP^{Sc}.[25] In a paper published in 2000, the domino model illustrated neuroinvasion in the peripheral nervous system. The authors put forward that there could exist 'a mode of transport in which PrP^{C} localized on the PNS is converted into PrP^{Sc} in a "domino" fashion centripetally towards the CNS' (Glatzel and Aguzzi, 2000: 2820), and then referred to their 1996 paper.

The progressive acceptance of the prion hypothesis was accompanied by other metaphors of graphic inspiration.[26] In textbooks or tutorials the 'rotten apple' metaphor is used to illustrate how a property can spread without information flow. For instance, on a website devoted to BSE one finds this statement: 'Like a rotten apple, once inside the brain, the mutant form of prion protein turns the native protein into more copies of the deviant, infectious form.'[27]

Work on the conformational change and the three-dimensional structure of prions led to a shift from the informational metaphor to these more graphic metaphors. In turn, the emergence of these new metaphors has contributed to further stimulate the search for the three-dimensional structure. As a result, since the mid-1990s knowledge of the three-dimensional structure has become a holy grail, and not only in Prusiner's work. As we will see below, many researchers have now joined this race and different approaches have been devised.

The race to the tertiary structure of prions: different means for a common goal

Over the years Prusiner's prion theory has gained widespread acceptance and computer technology was soon at the forefront of protein studies. Yet, if Prusiner's work has been decisive in establishing the prion theory, his papers on prion structure (mostly on the secondary structure) have not been regarded as seminal. Prusiner became only one contestant among others in the race to determine the three-dimensional structure of PrP.

In the opening speech of an international conference titled 'post-genomics', held at the Max Planck Institute for the History of Science in July 1998, its director H.-J. Rheinberger stressed that 'instead of being theory-driven, [molecular biology] appears to be eminently technology-driven'. This statement clearly applies to research on the structure of prions, where two main techniques have been used, namely crystallography (see note 7) and NMR.

Crystallography and NMR in prion research

Until the mid-1980s, the main approach to solve three-dimensional protein structure was through X-ray diffraction analysis, which makes use of crystallized proteins. Studies by John Kendrew and Max Perutz in the late 1950s, respectively on myoglobin and haemoglobin, were emblematic of the crystallographic approach. About ten years before, Felix Bloch and Edward Purcell had come up with the principle of NMR, which allows the detection of subatomic and structural information of molecules. In NMR a strong magnetic field (the stronger the field the higher the resolution) is applied to a sample and measures of how the system responds to radio waves are taken.[28] Initially NMR helped in chemistry to analyse quantitative mixtures containing known compounds. It took a long time for it to be applied to biological molecules.

William Dale Phillips (1925–93) was one of the pioneers in the late 1960s, when the Swiss Kurt Wüthrich arrived at the Bell Laboratories to work on NMR.[29] Wüthrich initially focused, as he recalls, 'on the metal ion coordination in the active sites of hemoproteins and on the electronic structure of the heme group' (Wüthrich, 2001: 923). It was only from the mid-1970s onwards that Wüthrich tried to apply NMR to *de novo* protein structures, that is, to proteins whose structure is unknown. In the late 1970s, Richard R. Ernst (1991 Nobel Prize for Chemistry) worked with Wüthrich to develop two-dimensional NMR techniques. In 1984, NMR proved as useful as X-ray crystallography to determine structures (Ottiger et al., 1994). In 2001, two representations of the backbones of the heavy atoms of a protein the tertiary structure of which was unknown (α-amylase inhibitor tendamistat), were independently produced with the two techniques and the graphic representations matched quite well (part b of Figure 4.12). This was regarded as visual confirmation that NMR was indeed of interest in molecular biology. In addition, NMR was a kind of complement to crystallography. Whereas crystallography supposed a fixed structure, NMR structure investigations were made in solution.[30] Wüthrich eventually received the 2002 Nobel Prize for Chemistry, 'for his development of nuclear magnetic resonance spectroscopy for determining the three-dimensional structure of biological macromolecules in solution'.[31]

Editor-in-chief of the *Journal of Biomolecular NMR*, Wüthrich is also best known for his application of NMR to the study of prions. In 1996, he proposed a solution for the tertiary structure of mouse PrP that conflicted with the structure given by Prusiner at that time.[32] Wüthrich found that mouse PrP^{C} (121–131) 'contains a two-stranded antiparallel beta-sheet and three alpha-helices' (Riek et al., 1996: 180).[33] More than

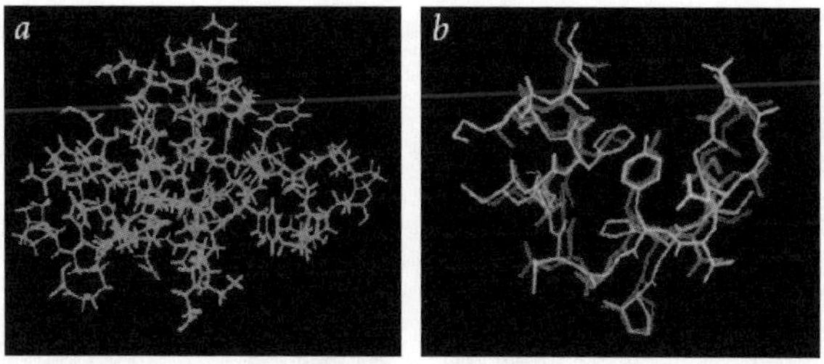

Figure 4.12 The first protein structure designed by NMR. The original colour figure can be found at http://prions.free.fr

Source: Wüthrich (2001). (Reprinted with permission from Kurt Wüthrich and from *Nature Structural Biology* 8: 924. © 2001 Nature Publishing Group.)

a third of his short paper was devoted to graphic representations and the range of representation modes was quite impressive (Figure 4.13). Crucially, a comparison with the mutation points identified in the primary structure of human PrP supported the proposed structure. From that time on NMR has been widely used in the study of prions and we have seen that Prusiner has also used it.[34] In a short history article published in 2001, Wüthrich confidently claimed: 'we may soon be able to obtain information on the structure of the disease-related, aggregated form of the prion protein' (Wüthrich, 2001: 925).

An important aspect of the 1996 paper on the structure of the mouse prion (Riek et al., 1996) was the use of bioinformatics. In the box where 'methods' were described, we read that 'the program MOLMOL was used to generate the figure'. The authors referred to a previous publication in the *Journal of Molecular Graphics* that described MOLMOL as a visualization device 'with special emphasis on nuclear magnetic resonance (NMR) solution structures of proteins' (Koradi et al., 1996: 51).[35] The classical approach of biochemistry had thus been complemented by structural biology, with its emphasis on 3D molecular structure.

In 1997, Wüthrich also used MOLMOL to design a monoclonal antibody that could be used to establish diagnosis (Corth et al., 1997). Three years later, in 2000, MOLMOL helped to visualize the human prion structure obtained by NMR (Zahn et al., 2000). A technical culture specific to computational biology is embodied in all software designed to represent structures. Progressively, this blurs the distinction between crystallographic and NMR methods: attention is paid to the structure *provided by visualization devices* irrespective of its mode of production. Moreover, whereas crystallography was dominant in the static approach, researchers have now started to use it for the identification of mobile parts of

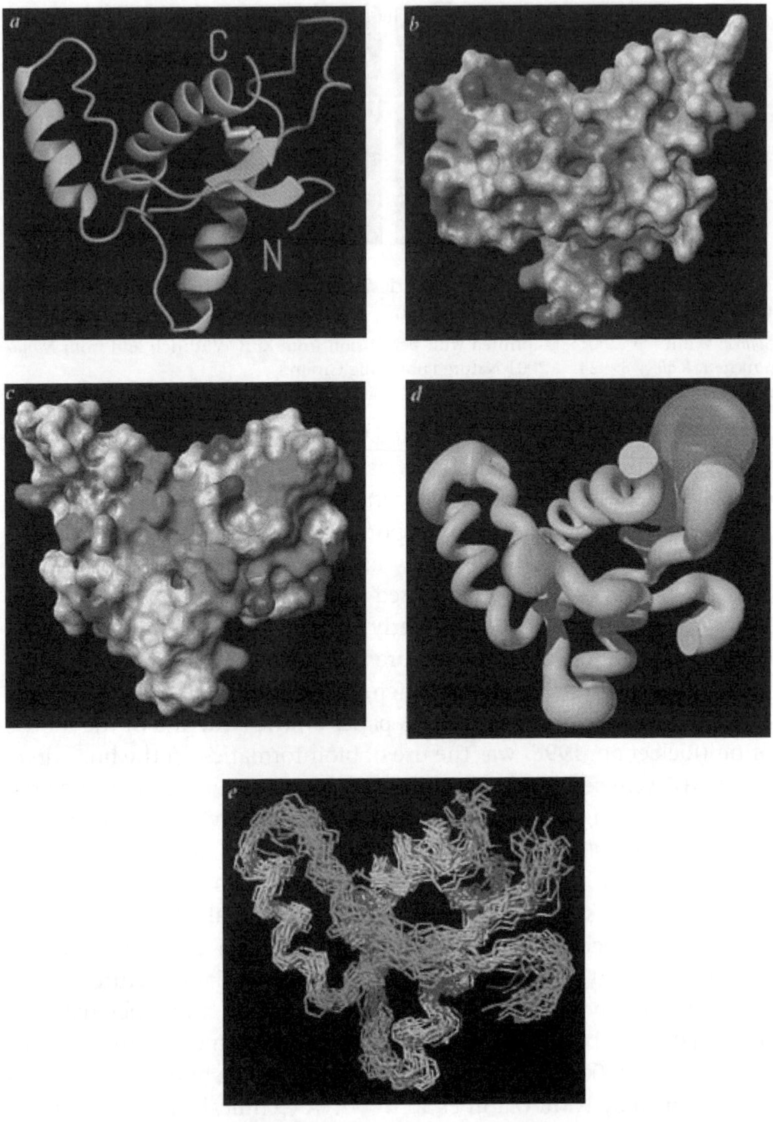

Figure 4.13 Tertiary structure of mouse PrP. The original colour figure can be found at http://prions.free.fr

Source: Riek et al. (1996). (Reprinted with permission from Kurt Wüthrich and from *Nature* 382: 181. © 1996 Nature Publishing Group.)

molecules. Though scientists still belong to one or the other scientific culture, this further blurs the difference between the two techniques.

Today, an image of the three-dimensional structure of human PrP^C stands on its own, with no caption, on Wüthrich's homepages.[36] The knowledge of PrP structure led to reification and just as the double helix stands for DNA, the structure of PrP^C now stands for the prion.

Salvation in the 'yeast prion'?

In the same way as the bacterium *Escherichia coli* served as the typical prokaryote organism (organism without nucleus) for the development of early molecular biology in the 1940s,[37] the yeast *Saccharomyces cerevisiae* has now become a useful organism for the development of a prion model. Yeast reproduces within a few hours and is thus much easier to handle than mammalian prions. Scientists assume that the understanding of the conformational change in yeast will provide valuable insight for studying mammalian PrP.

Since the mid-1980s a journal called *Yeast* has been devoted to these microorganisms. The *Yeast* editor for North America, Reed B. Wickner, whose background is in genetics, works at the National Institute of Health. In 1994, drawing upon studies conducted in the early 1970s by Cox and Lacroute,[38] Wickner was the first researcher to see in yeast an analogue to mammalian prions defined as infectious proteins. In an article published in *Science*, Wickner identified a 'prion analog', a protein that was found in two forms [Ure2], and [Ure3] for the altered form.[39] Wickner wrote in the abstract: 'In analogy to mammalian prions, [URE3] may be an altered form of Ure2p that is inactive for its normal function but can convert normal Ure2p to the altered form' (Wickner, 1994: 566).

Through the use of this analogy the prion concept came to be considered as a model rather than as a hypothesis.[40] The structure of yeast was studied in order to find 'prion-inducing domains' (Masison and Wickner, 1995). In a review article published in *Yeast* in 1995, Wickner explained that two yeast prions had been identified: [URE 3] and [PSI+], the altered form of the protein Sup35 (Wickner et al., 1995). The main work on [PSI+] was accomplished by Susan Lindquist, a specialist in heat-shock proteins (HSP, proteins produced when cells are exposed to warm temperatures, to ethanol, and other forms of environmental stress) (Chernoff et al., 1995). The concentration of an HSP named HSP104 determines whether Sup35 is 'cellular' or 'resistant', to use the notation introduced by Prusiner for mammalian prions.[41] Molecules like HSP, which can influence protein folding, are called 'chaperones', the theory of which was introduced in the mid-1980s.

Following the path-breaking work of Wickner and Lindquist, other scholars decided to work on yeast prions. One incentive was that security standards for yeast studies are not as high as those required for work on mammalian prions.[42] Young researchers like Ronald Melki and his team at the Laboratoire d'enzymologie et de chimie structurales (CNRS), took advantage of this aspect.[43] They decided to work on Ure2 gene products, expressed in *E. coli*. A geneticist working on the same site at Gif-sur-Yvette (France), Christophe Cullin, gave them the gene coding for Ure2p, and in 1999 this collaboration resulted in the publication of an important article in the *Journal of Biological Chemistry* (Thual et al., 1999). The authors characterized by biochemical methods Ure2p self-assembly in a dimeric state and gave 'strong evidence for the existence of at least two structural domains in Ure2p molecules' (Thual et al., 1999: 13666). They concluded their paper with a comparison with other 'prion-like' proteins.

In order to deepen the analysis of the three-dimensional structure, Melki and co-workers tried to crystallize Ure2, which had been purified to homogeneity. Since the characteristics of the beam line of the synchrotron they had access to were incompatible with the size of their crystals, they used the European Synchrotron Radiation Facility in Grenoble.[44] Their results were published in 2001 (Bousset et al., 2001). This was the first crystal structure (as opposed to 'NMR structure') of a protein with prion properties. A month later Wickner and colleagues also published a structure of Ure2 (Umland et al., 2001), though less complete than Melki's.[45] Many graphic representations were given in both papers: stereoviews of electron density maps, ribbon representations, and also comparison with already known structures like a bacterial glutathione S-transferase (GST). Crucially, both Melki and Wickner used the phrase 'prion protein' in the title of their papers, instead of 'prion-like' in Melki's 1999 paper (Thual et al., 1999).

The fact that Melki and Wickner arrived independently at a similar structure almost at the same time gives us an opportunity to see how the visualization cultures attached to crystallography (Melki) and to NMR (Wickner) have merged. Conventions have stabilized and biologists search in the same databases for proteins with similar structures. It is thus possible to compare their respective representations of the Ure2p dimer (Figure 4.14) and their respective superimpositions of this protein with *E. coli* or *Arabidopsis thaliana* GST (Figure 4.15).

The use of large-scale technical systems such as a synchrotron became embodied in representations and Melki's structure featured on the cover of the journal (Figure 4.16) that published his paper (Bousset et al.,

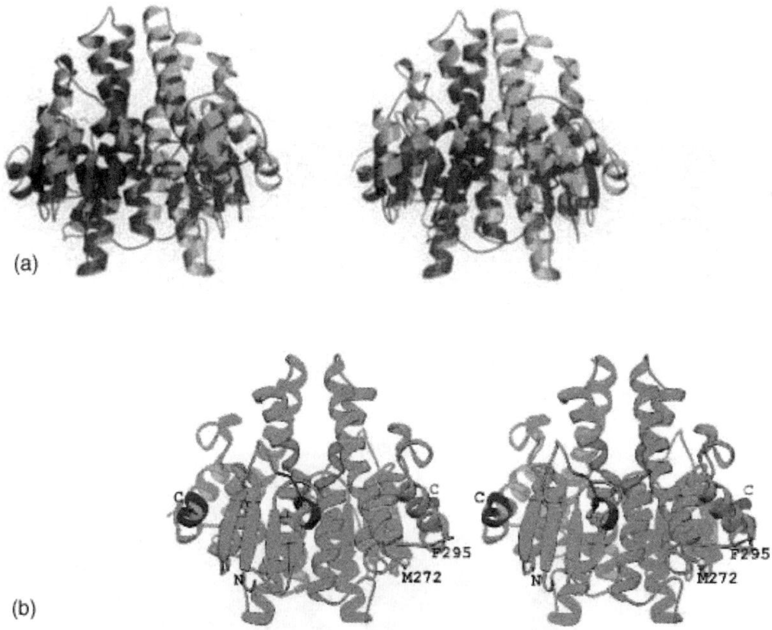

Figure 4.14 Stereoviews of a part of Ure2p (part 95-354 (a) and 97-354 (b)). The original colour figure can be found at http://prions.free.fr

Sources: (a) Bousset et al. (2001); (b) Umland et al. (2001). (Reprinted with both permissions from *C.R. Biologies* 325: 42 and *PNAs USA* 98(4): 1462. © 2001 Elsevier and 2001 National Academy of Sciences, USA.)

2001). At this point three-dimensional structures of prions, here of yeast, became emblematic of techniques, here of crystallography and the synchrotron. Figure 4.16 represented three dimers and the caption explained that this 'should help us to understand the mechanism of the amyloid formation associated with a number of degenerative diseases'. The underlying motivation of the race to the tertiary structure is to design drugs that can interfere with the structure to avoid aggregation. In this sense, graphic representations not only contributed to turning the prion hypothesis into a model, they also have a real heuristic power that should soon be appreciated.

This being said, if research on [URE 3] and [PSI+] in *Saccharomyces cerevisiae* has become paradigmatic of the prion model, some important considerations are sometimes missing from the modelization procedures. To begin with, no pathogenic effect has ever been noticed in yeast. One of the two phenotypes that have been studied presents

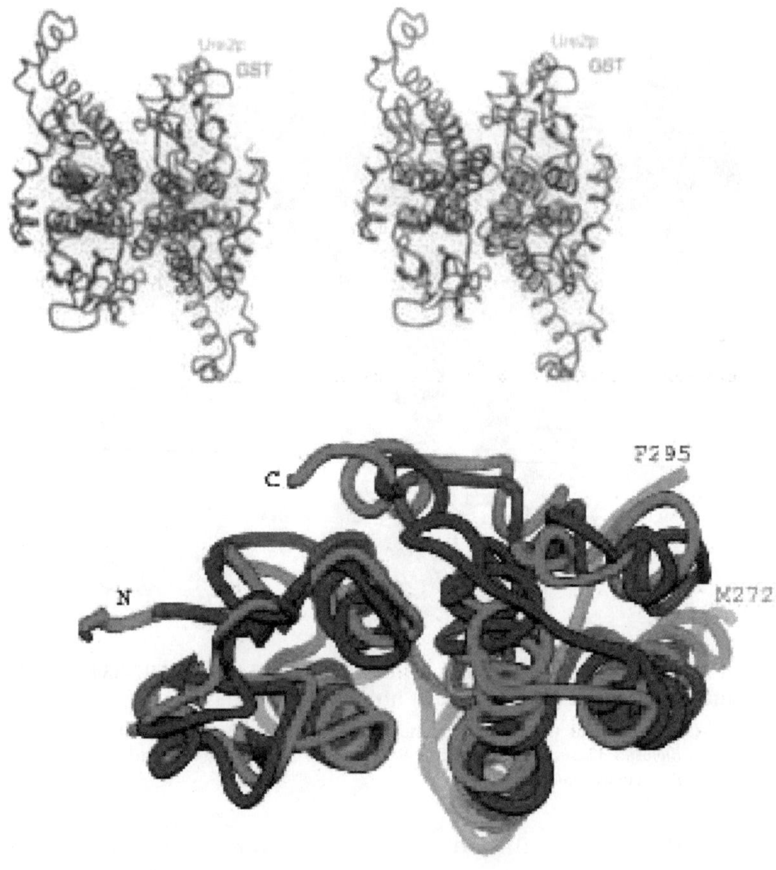

Figure 4.15 Stereoview of the superposition of the Ure2p dimer with *E. coli* GST (above) and superposition of a monomer of Ure2p with a monomer of *A. thaliana* GST (below). The original colour figure can be found at http://prions.free.fr

Sources: Bousset et al. (2001) and Umland et al. (2001). (Reprinted with both permissions from *C.R. Biologies* 325: 42 and *PNAS USA* 98(4): 1463. © 2001 Elsevier and 2001 National Academy of Sciences, USA.)

interesting aggregation properties similar to that of PrP^{Sc} but does not cause disease. If Ure2p can aggregate *in vitro*, it has now been proven that it does not cause [URE3] *in vivo*. Moreover, no homology sequence has been found between mammalian and yeast prions. To encapsulate these differences with mammalian prions, the term 'propagon' has recently been proposed to designate yeast prions.[46]

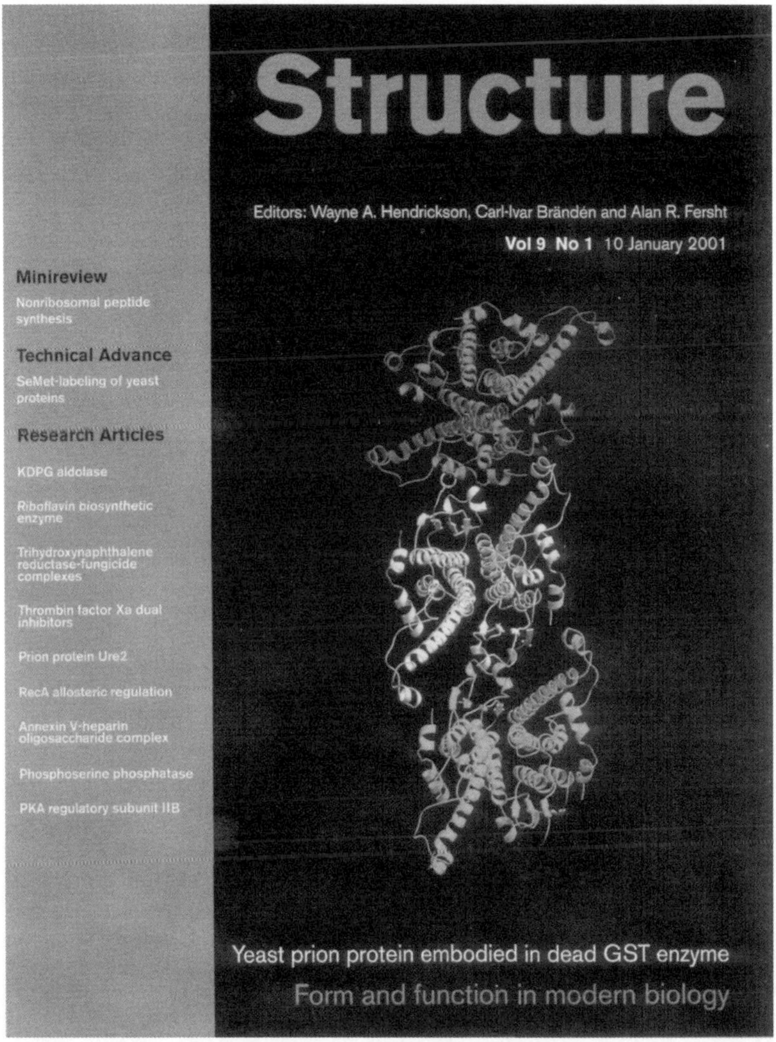

Figure 4.16 Structure of the yeast prion protein (*Structure*, front cover). The original colour figure can be found at http://prions.free.fr

Source: Structure (2001). (Reprinted with permission from *Structure* 91: front cover. © 2001 Elsevier.)

Conclusion

The present study has shown that in the early days the lack of representation of the three-dimensional structure of the prion protein, which is necessary to understand how it can convert into its pathogenic form, restricted the credibility of the prion hypothesis. As the French TSE expert Dominique Dormont put it: 'A scientific concept that is not supported by direct visualization is always difficult to establish, whatever its origin may be. In biology (I don't speak about physics or mathematics), something you cannot visualize always poses a lot of problems.'[47]

Yet, with the later work of Prusiner, Wüthrich, Wickner and Melki on prion structures, representations progressively became the core of the prion theory. In a recent paper on the 'Structure and assembly properties of the yeast prion Ure2', the word 'picture' has acquired a new meaning. The authors write that they hope to get a 'full picture of the molecular events at the origin of prion propagation' (Bousset et al., 2002: 6). Thanks to computer graphics allied to NMR or, in this case, crystallography, the epistemological function of visualization has moved from a mere illustration to a possible explanation of the very nature of prions.

Visualization has played an important role in changing the status of the prion concept from hypothesis to model. If biochemical experiments are still paramount, computer graphics have been essential to determine the three-dimensional structure of PrP^{C} and are likely to play a similar role in determining the structure of PrP^{Sc}. The development of therapeutics will conclusively establish the importance of three-dimensional structures since drug design aims at producing a molecule that can interfere with the structure of the pathological protein.

More generally, now that the Human Genome Project has been completed, there is little doubt that protein studies will continue to benefit greatly from the development of the prion hypothesis and its visualization culture.

Notes

1. The reluctance of philosophers and historians of science to pay attention to the visual aspects of scientific practice has already been discussed (Rudwick, 1976; Griesemer, 1991). Thomas Kuhn, for example, considered that visual representations were 'at best by-products of scientific activity' (Kuhn, 1977: 350).
2. For the most recent reviews of the science studies literature on visualization and visual representation, see Cambrosio (2000), Lynch (1998) and Soojung-Kim Pang (1997).

3. Latour and Woolgar (1979), Latour (1993), Amann and Knorr Cetina (1990) and Lynch (1985).
4. These secondary structures were first characterized by Linus Pauling and his colleagues in the early 1950s (Pauling et al., 1951).
5. An extensive literature deals with Crick's article. See Sarkar (1996) and Morange (1998).
6. For a rapid overview, see Levinthal (1966). Parallel to Levinthal's work, an 'Oak Ridge Thermal Ellipsoid Program' was also developed by Carroll Johnson (1965) to represent molecular structures using plot printers.
7. Crystallography is the X-ray analysis of the structure of crystallized proteins.
8. This work was achieved by David and Jane Richardson and colleagues, using a density-fitting computer system called 'GRIP' at the University of North Carolina. For an overview of the history of interactive molecular graphics, see Martz and Francoeur (2001).
9. On the notion of epistemic thing, see Rheinberger (1997).
10. On this subject, see Kim, Chapter 2 this volume.
11. This review article is largely based on a paper published in *Cell* (Prusiner et al., 1983).
12. In 1981, Patricia Merz at the Institute for Basic Research in Developmental Disabilities had already discovered that molecules in the central nervous system could have disease-related structures (Merz et al., 1981). She used electronic microscopy to isolate what she called 'scrapie-associated fibrils' (SAF), supposed to be pathogenic and similar to the 'rods' later shown by Prusiner. Merz later characterized SAF as a 'specific marker for the "unconventional" slow virus diseases' (Merz et al., 1984), whereas Prusiner used the same visualization device to promote his 'prion hypothesis'.
13. Compare with Figure 1.3 in Poulsen and Andersen's chapter, which represents four models given in another 1988 paper by the Prusiner group. Poulsen and Andersen (Chapter 1 this volume) also show how another diagram published in 1991 (their Figure 1.1) helped Prusiner to illustrate his theory.
14. For instance, 'structuralists', as they call themselves, questioned the use of Fourier transform infrared spectroscopy to compare the secondary structures of PrP^{C} and PrP^{Sc}. Interview with R. Melki, 18 February 2002, Paris.
15. See http://www.expasy.org/history.html or Appel et al. (1994).
16. For a description of NMR, see further in this chapter.
17. http://predictioncenter.llnl.gov/casp1/Casp1.html
18. Prusiner also used other programs, such as MidasPlus (Molecular Display and Simulation System), which were developed by the Computer Graphics Laboratory of the University of California where Prusiner's laboratory is located. On the Midas, see Ferrin et al. (1988).
19. On stereoviews, see Martz and Francoeur (2001).
20. Note that only three of the four helices predicted in 1994 were modelled. Prusiner remained silent about his false prediction.
21. The notion of chaperone molecules was introduced in the mid-1980s to denote molecules that influence proteins during their folding process. A paper was specifically devoted to the protein X hypothesis. See Kaneko et al. (1997).
22. See also the different documents available at http://www.nobel.se

23. On the role of two French biologists in the development of a kinetic approach (as opposed to Anfinsen's thermodynamical approach), see Segal (2002).
24. On the unconscious use of metaphors, see Lakoff and Johnson (1980).
25. This part of the paper was reproduced in Aguzzi et al. (1997) and Raeber et al. (1998).
26. On another level, see the 'Lufthansa route model' analysed by Dressel, Chapter 3 this volume, to explain inter-species transmission.
27. See http://www.mad-cow.org/~tom/prion_evol.html
28. For these findings, Bloch and Purcell were awarded the Nobel Prize for Physics in 1952.
29. On Phillips, see Shulamn (2000).
30. Wüthrich and other specialists are hoping that the recent development of solid-state NMR, first developed for short synthetic peptides, will allow them to find out the structure of aggregate proteins like PrP^{Sc}.
31. http://www.nobel.se/chemistry/laureates/2002/index.html
32. Wüthrich published a paper in TIBS which explicitly contradicted Prusiner's theoretical model. See Glockshuber et al. (1997).
33. The 1996 paper was complemented by another, Riek et al. (1997).
34. Prusiner's viewpoint on NMR analysis is described in Baldwin et al. (1998).
35. See also the official homepage at http://www.mol.biol.ethz.ch/wuthrich/software/molmol/. The *Journal of Molecular Graphics* published by the Molecular Graphics Society first appeared in 1983, marking a turning point in the institutionalization of the field.
36. See his two laboratories at http://www.mol.biol.ethz.ch/wuthrich/ and http://www.scripps.edu/mb/wuthrich/ (last accessed in February 2003).
37. A study has already been devoted to *E. coli* in cell differentiation of all species. See Thieffry (1996).
38. Non-Mendelian heredity in yeast was first noticed by B. S. Cox (Cox, 1965; Lacroute, 1971) and later by M. Aigle and F. Lacroute (Aigle and Lacroute, 1975). Because yeast cells are so easy to handle, it has been possible to establish in these organisms the new protein-based mechanism of heredity specific to the prion theory. The link between non-Mendelian heredity and the prion theory has been analysed by means of lexicography. See Maunoury et al. (1999).
39. As Wickner explains on his homepage, 'the normal function [of the URE2 protein] is to turn off utilization of poor nitrogen sources if a good nitrogen source is present' (http://www.ncbi.nlm.nih.gov/Yeast/wickner.html). The name 'Ure2' comes from ureidosuccinate because [URE3] is regarded as a non-Mendelian genetic element that makes cells able to take up ureidosuccinate when ammonia is the nitrogen source.
40. On the relevance of the yeast model, see Couzin (2002).
41. The resistant form analogous to PrP^{Sc} is noted $Sup35^{[PSI+]}$.
42. Whereas a P4 laboratory is necessary to experiment on mammalian prions, yeast work is done in a P3 laboratory since there are no risks of human contamination.
43. On the development of French research on prions in the early 1990s, see Chamak, Chapter 5 this volume.
44. The LURE was the synchrotron at their disposal (Laboratoire pour l'Utilisation du Rayonnement Electromagnétique).

45. A comparison of the two journals where these papers were published would be interesting. Members of the National Academy of Sciences like Wickner have the facility to publish in *PNAS* and the choice of journal is sometimes decisive in the diffusion of a scientific theory, which in turn depends on the referees ... .
46. On the differences between yeast and mammalian prions, see Fernandez-Bellot and Cullin (2001).
47. Interview with Dominique Dormont, 29 November 2001, Fontenay-aux-Roses. For more detail on Dormont's position in TSE research, see Chamak, Chapter 5 this volume.

References

Aguzzi, A. et al. (1997) 'Neuro-immune Connection in Spread of Prions in the Body?', *The Lancet* 349: 742–3.

Aigle, M. and F. Lacroute (1975) 'Genetical aspects of [URE3], a Non-mitochondrial, Cytoplasmically Inherited Mutation in Yeast', *Mol. Gen. Genet.* 136: 327–35.

Amann, K. and K. Knorr-Cetina (1990) 'The Fixation of (Visual) Evidence', in M. Lynch and S. Woolgar (eds) *Representation in Scientific Practice*. Cambridge: MIT Press: 85–122.

Anfinsen, C. (1973) 'Principles that Govern the Folding of Protein Chains', *Science* 181: 223–30.

Appel, R. D. et al. (1994) 'A New Generation of Information Retrieval Tools for Biologists: the Example of the ExPASy WWW Server', *Trends Biochem. Sci.* 19: 258–60.

Baldwin, M. A. et al. (1998) 'The Three-dimensional Structure of Prion Protein: Implications for Prion Disease', *Biochemical Society Transactions* 26: 481–6.

Bousset, L. et al. (2001) 'Structure of the Globular Region of the Prion Protein Ure2 from the Yeast Saccharomyces cerevisiae', *Structure* (Camb.) 9(1): 39–46.

Bousset, L. et al. (2002) 'Structure and Assembly Properties of the Yeast Prion Ure2', *C.R. Biologies* 325: 3–8.

Brandner, S. et al. (1996) 'Normal Host Prion Protein (PrP^{C}) is Required for Scrapie Spread within the Central Nervous System', *PNAS USA* 93: 13148–51.

Cambrosio, A. (2000) 'Argumentation, représentation, intervention: les rôles de l'imagerie dans les discours scientifiques', *ASp* 27/30: 95–112.

Chernoff, Y. O. et al. (1995) 'Role of the Chaperone Protein Hsp104 in Propagation of the Yeast Prion-like Factor [psi+]', *Science* 268(5212): 880–4.

Corth, C. et al. (1997) 'Prion (PrP^{Sc})-specific Epitope Defined by a Monoclonal Antibody', *Nature* 390(6655): 74–7.

Couzin, J. (2002) 'In Yeast, Prions' Killer Image Doesn't Apply', *Science* 297: 758–61.

Cox, B. S. (1965) 'PSI, a Cytoplasmic Suppressor of Super-suppressor in Yeast', *Heredity* 20: 505–21.

Crick, F. (1958) 'On Protein Synthesis', in F. K. Sanders (ed.) *The Biological Replication of Macromolecules*. Cambridge: Cambridge University Press: 138–63.

Donne, D. G. et al. (1997) 'Structure of the Recombinant Full-length Hamster Prion Protein PrP(29-231): the N Terminus is Highly Flexible', *PNAS USA* 94(25): 13452–7.

Fernandez-Bellot, E. and C. Cullin (2001) 'The Protein-only Theory and the Yeast Saccharomyces cerevisiae: the Prions and the Propagons', *Cell Mol. Life Sci.* 58(12–13): 1857–78.

Ferrin, T. E. et al. (1988) 'The MIDAS Display System', *J. Mol. Graphics* 6(1): 13–27, 36–7.

Francoeur, E. (2001) 'Molecular Models and the Articulation of Structural Constraints in Chemistry', in U. Klein (ed.) *Tools and Modes of Representation in the Laboratory Sciences*. Dordrecht: Kluwer Academic Publishers: 95–115.

Francoeur, E. and J. Segal (2004) 'From Model Kits to Interactive Computer Graphics', in S. de Chadarevian and N. Hopwood (eds) *Models: the Third Dimension of Science*. Stanford: Stanford University Press.

Gasset, M. et al. (1992) 'Predicted Alpha-helical Regions of the Prion Protein when Synthesized as Peptides Form Amyloid', *PNAS USA* 89(22): 10940–4.

Glatzel, M. and A. Aguzzi (2000) 'PrP^{C} Expression in the Peripheral Nervous System is a Determinant of Prion Neuroinvasion', *Journal of General Virology* 81: 2813–21.

Glockshuber, R. et al. (1997) 'Three-dimensional NMR Structure of a Self-folding Domain of the Prion Protein PrP(121–231)', *Trends in Biochemical Sciences* 22(7): 241–2.

Griesemer, J. R. (1991) 'Must Scientific Diagrams Be Eliminable? The Case of Path Analysis', *Biology and Philosophy* 6: 155–80.

Güntert, P. et al. (1991) 'Efficient Computation of Three-dimensional Protein Structures in Solution from Nuclear Magnetic Resonance Data Using the Program DIANA and the Supporting Programs CALIBA, HABAS and GLOMSA', *J. Mol. Biol.* 217: 517–30.

Huang, Z. et al. (1994) 'Proposed Three-dimensional Structure for the Cellular Prion Protein', *PNAS USA* 91(15): 7139–43.

Huang, Z. et al. (1996) 'Structures of Prion Proteins and Conformational Models for Prion Diseases', *Curr. Top. Microbiol. Immunol.* 207: 49–67.

James, T. L. et al. (1997) 'Solution Structure of a 142-residue Recombinant Prion Protein Corresponding to the Infectious Fragment of the Scrapie Isoform', *PNAS USA* 94(19): 10086–91.

Johnson, C. K. (1965) *OR TEP: a FORTRAN Thermal-ellipsoid Plot Program for Crystal Structure Illustrations*. ONRL Report No. 3794. Oak Ridge, Tenn.: Oak Ridge National Laboratory.

Kaneko, K. et al. (1997) 'Evidence for Protein X Binding to a Discontinuous Epitope on the Cellular Prion Protein during Scrapie Prion Propagation', *PNAS USA* 94: 10069–74.

Kay, L. E. (2000) *Who Wrote the Book of Life? A History of the Genetic Code*. Chicago: University of Chicago Press.

Keyes, M. E. (1999) 'The Prion Challenge to the "Central Dogma" of Molecular Biology, 1965–1991. Part I: Prelude to Prions' and 'Part II: The Problem with Prions', *Studies in History and Philosophy of Biological and Biomedical Sciences* 30 (1 and 2): 1–19, 181–218.

Koradi, R. et al. (1996) 'MOLMOL: a Program for Display and Analysis of Macromolecular Structures', *Journal of Molecular Graphics* 14(1): 51–5.

Kuhn, T. S. (1977) *The Essential Tension: Selected Studies in Scientific Tradition and Change*. Chicago: University of Chicago Press.

Lacroute, F. (1971) 'Non-Mendelian Mutation Allowing Ureidosuccinic Acid Uptake in Yeast', *J. Bacteriol.* 106: 519–22.

Lakoff, G. and M. Johnson (1980) *Metaphors We Live By*. Chicago: University of Chicago Press.

Latour, B. (1993) 'Le "Pédofil" De Boa Vista – Montage Photo-Philosophique', in B. Latour (ed.) *La Clef de Berlin*. Paris: La Découverte: 171–225.

Latour, B. and S. Woolgar (1979) *Laboratory Life: the Social Construction of Scientific Facts*. Beverly Hills: Sage Publications.

Levinthal, C. (1966) 'Molecular Model-Building by Computers', *Scientific American* 214: 42–52.

Lynch, M. (1985) 'Discipline and the Material Form of Images', *Social Studies of Science* 15(1): 37–66.

Lynch, M. (1998) 'The Production of Scientific Images: Vision and Re-Vision in the History, Philosophy, and Sociology of Science', *Communication and Cognition* 31(2–3): 213–28.

Martz, E. and E. Francoeur (2001) 'History of Visualization of Biological Macromolecules', http://www.umass.edu/microbio/rasmol/history.htm (latest revision 9/2001).

Masison, D. C. and R. B. Wickner (1995) 'Prion-inducing Domain of Yeast Ure2p and Protease Resistance of Ure2p in Prion-containing Cells', *Science* 270(5233): 93–5.

Maunoury, M. T. et al. (1999) 'Observer la science en action ou, comment les sciences de l'information permettent de suivre l'évolution et la convergence des concepts de prion et d'hérédité non mendélienne dans la littérature', *Médecine/Sciences* 15(4): 577–82.

Merz, P. et al. (1981) 'Abnormal Fibrils from Scrapie-infected Brain', *Acta Neuropathologica* 54: 63–74.

Merz, P. et al. (1984) 'Infection-Specific Particle from the Unconventional Slow Virus Diseases', *Science* 225: 437–40.

Morange, M. (1998) *A History of Molecular Biology*. Cambridge: Harvard University Press.

Ottiger, M. et al. (1994) 'The NMR Solution Structure of the Pheromone Er-2 from the Ciliated Protozoan Euplotes Raikovi', *Protein Science* 3: 1515–26.

Pauling, L. et al. (1951) 'The Structure of Proteins: Two Hydrogen-Bonded Helical Configurations of the Polypeptide Chain', *PNAS USA* 37: 205–11.

Prusiner, S. (1982) 'Novel Proteinaceous Infectious Particles Cause Scrapie', *Science* 216: 136–44.

Prusiner, S. (1984) 'Prions', *Scientific American* 251: 48–57.

Prusiner, S. (1988) 'Molecular Structure, Biology, and Genetics of Prions', *Advances in Virus Research* 35: 83–136.

Prusiner, S. (1992) 'Chemistry and Biology of Prions', *Biochemistry* 31(49): 12277–88.

Prusiner, S. (1994) 'Biology and Genetics of Prion Diseases', *Annu. Rev. Microbiol.* 48: 655–86.

Prusiner, S. (1998) 'Prions', *PNAS USA* 95(23): 13363–83.

Prusiner, S. et al. (1983) 'Scrapie Prions Aggregate to Form Amyloid-like Birefringent Rods', *Cell* 35: 349–58.

Prusiner, S. et al. (1987) 'On the Biology of Prions', *Acta Neuropathologica* 72: 299–314.

Raeber, A. J. et al. (1998) 'Transgenic and Knockout Mice in Research on Prion Diseases', *Brain Pathology* 8: 715–33.

Rheinberger, H. J. (1997) *Toward a History of Epistemic Things: Synthesizing Proteins in the Test Tube (Writing Science)*. Stanford: Stanford University Press.

Riek, R. et al. (1996) 'NMR Structure of the Mouse Prion Protein PrP(121-231)', *Nature* 382: 180–2.

Riek, R. et al. (1997) 'NMR Characterization of the Full-length Recombinant Murine Prion Protein mPrP (23-231)', *FEBS Letters* 413: 282–8.

Rudwick, M. J. S. (1976) 'The Emergence of a Visual Language for Geological Science', *History of Science* 14: 149–95.

Sarkar, S. S. (1996) 'Thinking of Biology – Decoding "Coding" – Information and DNA', *Bioscience* 46: 857–64.

Segal, J. (2002) 'Les premiers "replieurs" français: Michel Goldberg à l'Institut Pasteur et Jeannine Yon à Orsay', *Revue pour l'histoire du CNRS* 7: 50–6.

Shulamn, R. G. (2000) 'D. C. Phillips', *Biographical Memoirs of the National Academy of Sciences*. National Academy Press: 166–81 (or http://books.nap.edu/books/030907035X/html/166.html)

Soojung-Kim Pang, A. (1997) 'Visual Representation and Post-Constructivist History of Science', *Historical Studies in the Physical and Biological Sciences* 28: 139–71.

Thieffry, D. (1996) 'E. coli as a Model System with which to Study Cell-differentiation', *Hist. Philos. Life Sci.* 18: 163–93.

Thual, C. et al. (1999) 'Structural Characterization of Saccharomyces cerevisiae Prion-like Protein Ure2', *J. Biol. Chem.* 274(19): 13666–74.

Umland, T. C. et al. (2001) 'The Crystal Structure of the Nitrogen Regulation Fragment of the Yeast Prion Protein Ure2p', *PNAS USA* 98(4): 1459–64.

Wickner, R. B. (1994) '[URE3] as an Altered URE2 Protein: Evidence for a Prion Analog in Saccharomyces cerevisiae', *Science* 264(5158): 566–9.

Wickner, R. B. et al. (1995) '[PSI] and [URE3] as Yeast Prions', *Yeast* 11(16): 1671–85.

Wüthrich, K. (2001) The Way to NMR Structures of Proteins', *Nature Structural Biology* 8: 923–5.

Zahn, R. et al. (2000) 'NMR Solution Structure of the Human Prion Protein', *PNAS USA* 97(1): 145–50.

5
Prion Research and the Public Sphere in France

Brigitte Chamak

Introduction

French TSE scientists have carried out pioneering studies but their work is rarely mentioned by commentators. In particular, the work of French radiobiologist Raymond Latarjet (1911–98) who collaborated with Tikvah Alper (1909–95) is almost unknown. The aim of the present chapter is therefore to outline the history of TSE research in France. I will explore the constitution of the TSE research network and the reception of the prion hypothesis in France with a view to demonstrating that the growth of scientific fields and the settlement of scientific controversies are often linked to developments in the public sphere. In France, actors located outside the scientific community were heavily involved in the expansion of TSE research and in the apparent resolution of the prion controversy.

To analyse TSE research in France I will draw upon the distinction between prion research and prion hypothesis. The *prion hypothesis* holds that the prion protein PrP can be infectious and cause TSEs. It has brought about a controversy in the TSE field because it postulates a novel mechanism of infection that differs from stabilized knowledge on pathogens. Since it is still awaiting final confirmation some scientists doubt that it is an accurate explanation of TSE aetiology. This being said, when Prusiner and colleagues succeeded in isolating the prion protein gene (Oesch et al., 1985), they opened up a new research programme in the TSE field – prion research, which offered scientists several new lines of inquiry and was, in this respect, very successful. *Prion research* is comprised of the numerous studies that look into the role of the prion protein or prion gene in TSEs. We will see below that there is no simple correspondence between participation in prion research and endorsement of the prion hypothesis.

The present study draws upon a range of sources: interviews with French TSE experts, public reports, publications on TSEs, and the archives of the Army Health Service (AHS). In the first part of the chapter I will describe French TSE research from the early days until 1996. Questions such as the low number of French TSE laboratories and the involvement of physicians from the AHS are addressed. In 1992, only six laboratories in France were working on TSEs. However, following the iatrogenic contamination of patients treated with growth hormone, the numerous cases of bovine spongiform encephalopathy (BSE) in the UK, the occurrence of some cases in France and the possible transmission of the disease to humans, a report was commissioned in 1992 by the French Minister of Research. The analysis of this report allows us to map the network of French experts and to assess the reception of the prion hypothesis at that time. In the second part I will deal with the implications for French TSE research of the March 1996 British announcement on the causal link between BSE in cows and vCJD in humans. While many new French TSE workers were led into prion research, at that time resistance to the prion hypothesis was quite strong. In the third part of this chapter I will review the current situation in the field – a situation which is characterized by the dominance of the prion hypothesis. I will also describe press coverage of TSEs since the beginning of the 1990s, and will show how the media's duty to offer an explanation of these diseases has now led them to back the prion hypothesis.

Part I. From the early days to 1996

The first TSE researchers

Books and articles on the history of TSE research focus on the work of Carleton Gajdusek, Stanley Prusiner and British and Scottish groups, yet French research is seldom mentioned (Beauvais and Billette de Villemeur, 1996; Rhodes, 1997; Brown and Bradley, 1998; Collinge, 2001; Schwartz, 2001). Only the Toulouse school of veterinary medicine is sometimes referred to, in particular the experiments of Jean Cuillé and Paul-Louis Chelle in the 1930s. These veterinarians demonstrated for the first time that scrapie was a transmissible disease (Cuillé and Chelle, 1936). These experiments ended during the Second World War, mainly because of Chelle's death and Cuillé's retirement (Barrairon, 1990).

In 1970, Raymond Latarjet, a radiobiologist at the Curie Institute, published with Tikvah Alper's group an article in *Nature* which showed that the scrapie agent resisted huge doses of ultraviolet light at a wavelength specifically absorbed by nucleic acid (Latarjet et al., 1970). In their introduction, Latarjet and colleagues explained that because scrapie is transmitted by cell-free filtrates the agent has been classified as a virus, though its response to many chemical and physical treatments has long been known to differ from that of 'conventional' viruses. However, Latarjet did not pursue this type of research. In the 1970s, it was considered to be of little economic importance[1] (Hornsey and Denekamp, 1997). Funding for research on human TSEs, considered as rare diseases, was also limited.

At the time Latarjet published his work with Alper's group, Louis Court was a research assistant at the AHS[2] and was also Latarjet's student. He was impressed by their results on the scrapie agent. The topic of his thesis, submitted in 1968, was the effects of radiation on the central nervous system. He subsequently carried out experiments with monkeys. He was soon the best specialist in electroencephalography with monkeys.[3] In 1970, Court was contacted by neurologist Françoise Cathala to devise an early diagnostic test for Creutzfeldt–Jakob disease (CJD) which recorded electroencephalographic signals in the brain of chimpanzees injected with extracts of infected material. Abnormal electroencephalographic patterns were observed before the onset of clinical signs. However, these results were not published because they were not considered specific enough.[4]

From the 1960s, Cathala was interested in the persistence of viruses in the central nervous system; however, she failed to find academic support in France.[5] In 1968, after reading the papers published by Gajdusek's group – in particular the one that first described kuru (Gajdusek and Zigas, 1957) – Cathala decided to join Gajdusek's laboratory in Bethesda. She studied brain biopsies and participated in various activities of the laboratory. Cathala was then employed by INSERM (Institut National de la Santé et de la Recherche Médicale). She returned to France with the intention of reproducing the experiments of kuru transmission to monkeys. Unfortunately, such experiments were expensive and required special equipment. She was unable to continue this type of research and decided to undertake epidemiological studies of CJD with the epidemiologist Paul Brown from Gajdusek's laboratory. Cathala was in contact with nearly all the French neurologists and thus succeeded in working on CJD cases from all regions of France (Brown et al., 1987). She also studied the distribution of scrapie in France and its possible transmission to humans (Cathala et al., 1985). The importance of her work was not appreciated and she did not receive sufficient funding. Cathala retired in the early 1980s. She donated to Court the equipment she had acquired at the Salpêtrière hospital where she was based.

Court was heavily involved in the organization of the first international symposium on unconventional virus diseases of the central nervous system, held in Paris in 1981. At this symposium Stanley Prusiner presented for the first time his prion hypothesis (Court, 1983). The second congress (1986), and the third (18–20 March 1996), both held at the Val de Grâce hospital in Paris, were also organized by Court.

Court had a student, Dominique Dormont, who is currently the spokesman of French TSE research. In the late 1970s, Dormont studied scrapie and the production of phosphatases.[6] In 1982, he moved to Gajdusek's laboratory. When he returned to France, in January 1983, Françoise Barré, in Luc Montagnier's laboratory at the Pasteur Institute, isolated HIV. Dormont worked with her on the isolation of reverse transcriptase. He was interested in the hypothesis that TSEs were caused by a retrovirus and tried to back it by searching for reverse transcriptase and DNA polymerase anomalies in spleens of TSE-affected animals. Though modification of DNA polymerase activity was observed, it was interpreted as a consequence of a degenerative process. In 1984, Dormont gave up this work but went on to study AIDS, TSEs and neurodegenerative processes.[7] He applied to TSEs the concepts, experimental protocols and tools used to study viral diseases.

Dormont's collaboration with virologists seems paradoxical: the intellectual filiation between Latarjet and Dormont (through Court) should have 'logically' led him to take up the prion hypothesis at an early stage. Far from that, he chose to subscribe to the viral explanation and still does. Three factors account for this surprising aspect. Firstly, Latarjet and Alper's results could actually be construed in at least two different ways: either the existence of an infectious protein, or an unconventional virus masked by a protein. Secondly, since Court and Dormont were researchers at the AHS and funded by the military, the second, more orthodox interpretation, was probably more acceptable to them. Thirdly, Dormont and other French scientists collaborated with Gajdusek, who, for years, regarded TSEs as viral diseases. Neither Cathala, nor Court and Dormont had links with Prusiner's group. Cathala was interested in the persistence of viruses in the central nervous system, Court conducted research on the epidemiology and biology of encephalopathies due to 'unconventional viruses',[8] and Dormont worked with HIV experts and tried to demonstrate the involvement of a virus in TSEs, borrowing tools and techniques used in HIV research.[9]

This brief account shows that far from being institutionally organized, early TSE research in France was conditional upon individual initiatives. A few researchers working on other topics managed to devote some of their funding and time to TSEs. In particular, as a radiation expert Court succeeded in obtaining funds from the DRET (Direction des Recherches et Etudes Techniques).[10] In an interview he explained that the military were interested in the protection against radiation and especially in the TSE agent's resistance to them.[11] At the AHS, TSE research was regarded as fundamental but not as a priority (Court, 1996). For his part, Dormont pursued TSE research but his group worked mainly on HIV. In short, no structural support existed for TSE research. We will see below that political developments later changed this state of affairs.

The 1992 report

In 1985, 1987 and 1989, Dormont, who was by then one of the few French TSE experts, was contacted by the General Health Services (Direction Générale de la Santé) to produce reports on the iatrogenic contamination of patients treated with human growth hormone (GH). In 1985, the identification of CJD cases following GH injections led to the suspension of GH treatments in the USA and other countries, but not in France where no suspect case had been identified.

In the early 1980s, Annick Alperovitch, a French epidemiologist, was nominated by the INSERM to conduct epidemiological studies on children treated with GH. The first cases of CJD appeared in 1989, followed by 20 cases in 1993, 40 in 1996, and 81 in 2001. The French national programme of GH treatment, initiated by Professor Royer, was set up in 1974 and involved the France-Hypophyse Association and the Pasteur Institute for production, and the Central Pharmacy of Paris hospitals for conditioning and supply (Chamak, 1999). An investigation directed by the Central Inspectorate of Social Affairs, published in 1992, revealed that not enough precautions were taken during the collection and production of GH (Clement et al., 1992).

The GH affair, along with the BSE epidemic in the UK, later led the Minister of Research Hubert Curien to commission a report on TSEs with a view to improving decision-making. On 6 April 1992, Curien wrote to Dormont and asked him to review existing knowledge on TSEs and to make recommendations for further development of TSE research. The 1992 report was the first of a series of reports on TSEs commissioned by the French Ministry of Research (Dormont, 1992). This report is invaluable since it allows us to map the TSE research network and to assess the status of the prion hypothesis in France at that time.

Network of TSE researchers

The report reveals that by 1992 only six research centres in France were working on TSEs:

- the laboratory of neuropathology and neurovirology of the AHS and CEA (Centre d'Énergie Atomique) in Fontenay aux Roses;
- the laboratory of neurochemistry at Saint-Louis hospital in Paris;
- the Pasteur Institute in Lille;
- the laboratory specializing in bovine pathologies at the CNEVA (Centre National d'Études Vétérinaires et Alimentaires) in Lyon;
- the laboratory of neuropathology at the Salpêtrière hospital in Paris;
- the INSERM epidemiological unit (U 360) in Villejuif.

Only a small number of researchers and technicians were dedicated to TSE research: 9 full-time researchers, 14 researchers with additional activities, and around 10 technicians. In the CEA laboratory, Dominique Dormont, Louis Court, Corinne Lasmézas (veterinarian) and Jean-Philippe Deslys (physician) studied the molecular physiopathology and the early molecular markers of TSEs. In Saint-Louis hospital, Jean-Louis Laplanche and Jacqueline Chatelain tried to identify changes in the PrP gene in CJD and scrapie. Philippe Amouyel, Camille Locht and Jean-Yves Cesbron, at the Pasteur Institute in Lille, studied modifications of the PrP gene and the species barrier. In Lyon, the laboratory headed by Marc Savey carried out an epidemiological survey of BSE. Jean-Jacques Hauw from the Salpêtrière hospital became an expert in human TSE diagnosis. The INSERM unit headed by Annick Alpérovitch was dedicated to the epidemiological survey of CJD.

All these researchers contributed to the 1992 report. They complained that public research bodies (INSERM, CNRS, INRA) and the Pasteur Institute were not developing specific structures adapted to the problems raised by TSEs. Only individual initiatives from researchers of these institutions, funding from the DMA, and collaboration with the National Institute of Health in the USA gave rise to studies in this area.[12]

In the report two risks for human health were identified. First, the use of biological material of human origin (grafts, biomaterials, drugs, cosmetology). Second, the potential transmission of BSE to humans via food products and therapeutic products of bovine origin (biomaterials, biotechnology and cosmetology).[13]

Status of the prion hypothesis

The 1992 report also contains a discussion on the nature of TSE agents.[14] With the exception of Laplanche,[15] at that time French researchers did not collaborate with Prusiner. The absence of links with Prusiner's group was understandable: at that time a French network already existed with a high level of knowledge on TSEs. Court and Cathala were working with Gajdusek even before Prusiner started to study scrapie. As a result, French workers did not regard Prusiner as the 'TSE Master'. Thus, the 1992 report explained:[16]

> The scientific community now agrees to retain the major role of PrP in these human and animal diseases, and to consider the accumulation of PrP^{Sc} as a specific criterion of TSEs. On the other hand a number of disagreements persist concerning the exact role of this protein: is it the infectious agent itself or one of its components? Or maybe PrP^{Sc} is only the reactive neuropathological product due to the proliferation of the agent and/or the marker of neuronal degeneration?

The report also emphasized the importance of Hsiao and colleagues' work who, in 1989, showed a link between a mutation of the PrP gene at codon 102 and a form of Gerstmann–Sträussler–Scheinker syndrome (GSS).[17] However, authors of the report also mentioned that no mutation was found in several families with GSS and CJD, suggesting that the PrP gene was not the only gene involved in TSEs.[18] The importance of the polymorphism at codon 129 of the PrP gene was also stressed because homozygosis at this level appeared as a genetic predisposition to develop CJD. The authors pointed out that the hypothesis of a genetic predisposition to develop a TSE under the influence of exogenous factors could not be excluded.[19]

Thus, in the 1992 report the prion hypothesis was presented but was not considered as the only possible answer to the scientific questions raised by TSEs. Moreover, the authors stressed that in the prion hypothesis initially formulated by Prusiner in 1982, the notion of a self-replicating foreign protein had been postulated and had now been abandoned.[20]

The 'virino' hypothesis of Alan Dickinson's group was also presented in the report:[21] 'These agents would comprise genetic information but would be surrounded by proteolipidic molecules belonging to the host explaining the lack of immune response.'

The report also introduced Weissmann's 'holoprion' hypothesis (Weissmann, 1991). In this model, PrP^{Sc}, called 'apoprion', would be responsible for the neuropathological phenomena whereas infectivity and the existence of different strains could be due to a nucleic acid, called 'coprion', of an unknown nature. This coprion could be dependent or not upon host genome. The association apoprion–coprion would be the holoprion. This hypothesis combines the virino hypothesis and the results obtained from research on PrP.

The conclusion of the 1992 report was that any of these hypotheses could account for clinical, biological and epidemiological data.[22]

Even though the prion hypothesis was regarded as only one possible explanation among others, prion research was developing in France: a growing number of studies on PrP protein and PrP gene were undertaken. French researchers had no doubt concerning the involvement of PrP but debated the real nature of the transmissible TSE agent.

Whatever the nature of the agent, the risk of transmission was considered to be high but, at the same time, authors of the 1992 report claimed that the risk of BSE transmission to humans was probably low.[23] They nevertheless recommended carrying out an epidemiological survey and imposing controls on industry to ascertain the origin of their products, especially bovine products.[24] The report recognized that the iatrogenic transmission of CJD by GH treatment on the one hand, and the emergence of BSE on the other, posed a major problem both for human and animal health. Consequently, it insisted on the importance of understanding the nature of the infectious agent, and recommended sustained research in this field.[25] However, the 1992 report came out of the personal initiative of the Minister of Research, who did not act at the government's request. For the French authorities the problems raised by GH contamination and BSE appeared rather limited and this perception was not counterbalanced by sufficient media pressure. As a result, funding was postponed until April 1996, just after the British announcement on the probable transmission of BSE to humans.

Part II. The turning point: the BSE crisis of March 1996

The TSE Expert Committee and the research programme

On 20 March 1996, the British government announced that ten cases of a new variant of CJD (vCJD) had been identified by the CJD Surveillance Unit, and claimed that this represented BSE transmission to humans. This announcement provoked a major European crisis at the economic, political and public health levels. This crisis was particularly acute in France. Confidence in governmental risk communication was severely undermined (Petitjean, 1996; Hirsch et al., 1996; Brizay et al., 1997; Mattéi, 1997; Chamak, 1999). The UK exported meat-and-bone meal, thought to be the cause of the BSE epidemic, and BSE cases had been found in France. Hence, the possible emergence of cases of vCJD could not be ruled out. The decrease in meat consumption was detrimental to the Ministry of Agriculture and to farmers. In response to the crisis the French government decided to support TSE research. The aims were to restore public confidence, to improve decision-making for the protection of human and animal health, and to develop diagnostic methods for the early detection of BSE.

A national TSE research programme was set up under the auspices of the Ministry of Research, Ministry of Agriculture and Secretary of State for Health. With 1.6 million French francs (MF), the programme's budget was substantial. An Expert Committee headed by Dominique Dormont was set up on 17 April 1996. The committee was comprised of those scientists who contributed to the 1992 report and of new actors. In all, 27 experts sat on the committee. Nearly half of them were veterinarians, one-third were biologists, and the others were neurologists or epidemiologists. Several agencies and institutes were represented: CEA, CNRS (Centre National de la Recherche Scientifique), INSERM, INRA (Institut National de la Recherche Agronomique), ENV (École Nationale Vétérinaire d'Alfort), CNEVA and the Pasteur Institute.

The Expert Committee's remit was to update knowledge of TSEs, to contribute to decision-making, and to devise a research programme in conjunction with European partners. The committee organized workshops to spread information on occupational risks. Collaboration with British researchers was established to exchange material and data. Another role of the committee was to answer questions posed by senior officials. With respect to the research programme, the committee established the lines of inquiry that should be developed and, after peer review, selected the projects that were to receive funding. Biology of TSE agents, physiopathology, sociology and risk analysis, safety and therapeutics, biological tools and new diagnostic methods were the themes chosen. The programme was coordinated by INSERM and the CEA was also strongly involved.

In the 1990s, prion research was flourishing in the USA. Prusiner's prestige was consolidating and peer review was increasingly in favour of the prion hypothesis. As a result, the 1996 French programme led newcomers to the field of prion work, especially young scientists often enthusiastic about the prion hypothesis. Important findings were produced. Joëlle Chabry, INSERM researcher in a CNRS laboratory (Nice), isolated a peptide that is part of the PrP protein and can prevent interaction between the normal and the pathological isoforms of the protein (Chabry et al., 1998, 1999). Sylvain Lehmann, INSERM researcher in a CNRS laboratory (Montpellier) who has worked in Prusiner's laboratory, studied the functions of PrP in an *in vitro* model. He developed a permanent cell line sustaining the propagation of natural sheep scrapie and over-expression of ovine PrP (Vilette et al., 2001). Working on the immune aspects of TSEs, Paul Aucouturier at INSERM in Paris found that infection of splenic dendritic cells with scrapie was sufficient for prion transmission to the CNS in mice (Aucouturier et al., 2001). Bernard Rossi and his colleagues in Nice developed a method to create chimeric prion proteins, using PrP and immunoglobulin fragments that can be detected by the immune system. Michel Laurent and his collaborators at the CNRS studied the dynamics of infection and properties of the transition from the normal to the pathological isoform of PrP (Kellershohn and Laurent, 2001). Another type of prion-related research was done in yeast by Cristophe Cullin and Ronald Melki at the CNRS (Komar et al., 1997, 1999; Thual et al., 1999, 2001).

It should be stressed that the above researchers did not necessarily believe in the prion hypothesis. However, all were trying to understand the action of prions and to find out whether or not the prion protein was indeed the infectious agent.

Other workers were more loosely involved in prion research. Michel Dron, INSERM researcher in a CNRS laboratory (Villejuif) looked for an increased expression of genes after infection and discovered that one such gene codes for an unknown protein called SRG1 (scrapie responsive gene 1) (Dron et al., 2000). Jean-Paul Fuchs, CNRS researcher in an INSERM laboratory (Strasbourg), set up a project aimed at comparing the expression of proteins in brains of TSE-affected mice and in vCJD-affected human brain. The objective was to produce evidence that other proteins were involved in these pathologies (Rangon et al., 2003). This type of work was compatible with prion research and even with the prion hypothesis since Prusiner himself has claimed that a protein other than PrP is involved in prion propagation (Telling et al., 1995).

However, work that openly conflicted with the prion hypothesis was somewhat hindered, even if it addressed the role of PrP. Jean-Luc Darlix at INSERM (Lyon) explored other hypotheses than that advanced by Prusiner. Darlix found that the prion protein mimics a viral protein GAG, and that *in vitro* prion protein binds large nucleic acids causing formation of nucleoprotein complexes resembling human type 1 immunodeficiency virus nucleocapsid–RNA complexes (Gabus et al., 2001a, b). In a model of HIV, when Darlix and his collaborators tried to over-express the prion protein, they succeeded in demonstrating that HIV replication decreased. They suggested that the prion protein could thus play a role in the reactions against the viral infection and, when a huge infection occurs, the binding of the prion protein to the viral nucleic acid could change its form and make it resistant to enzymatic degradation. These results suggested that viral infection precedes the prion reaction. In other words, they clashed with the prion hypothesis. Darlix experienced difficulties in getting his work published, in establishing collaboration with other workers, and even had problems in obtaining funding in 1996. In an interview he said that challenging the mainstream view requires the challenger to produce much more data and more arguments than those who tend to confirm it. He also put forward that being predominantly Anglo-American, the referee system works against French researchers.[26]

This brief review of projects carried out under the first TSE research programme indicates that researchers' theoretical framework, either prion or virus, guided their practices, experimental approach and results. Importantly, at that time Dormont's position of power in the French landscape contributed to keeping the virus school alive despite the development of prion research and the growing influence of the prion hypothesis. First, Dormont headed a laboratory where young researchers were, and still are, trained along the lines of the viral hypothesis.[27] Second, he produced several reports for the French government in which the prion hypothesis was questioned. Third, he headed the Expert Committee that selected the projects that were to get funding, and since 1996 his own laboratory has received substantial funds. Finally, regarded as *the* TSE expert in France, Dormont was constantly called in by the media. We will see below that with the second programme set up in November 2000, his ability to influence the orientation of French research weakened.

Part III. The current situation of the TSE field

TSE research

We have seen that French authorities took the BSE crisis of March 1996 very seriously. This tendency has clearly been sustained as demonstrated by the massive increase in TSE funding (Table 5.1).

Table 5.1 TSE research funding in France, 1996–2001

1996	*1997*	*1998*	*1999*	*2000*	*2001*
1.6 MF	6.7 MF	8.8 MF	15 MF	70 MF	160 MF

Since 1999 a number of additional measures have been taken by the French government. In 1999, the AFSSA (Agence Française de Sécurité Sanitaire des Aliments) was set up. Its remit is to assess food-related risks, to conduct research and to provide expert advice. Just as the creation of the UK Food Standards Agency, this was a political decision taken by politicians confronted with public distrust about food safety.

In 2001, a new laboratory dedicated to research on TSE diagnosis was inaugurated at the CEA (where Dormont is based), which received more than 10 MF from the Ministry of Research. This was part of a broader funding strategy. In November 2000, the Minister of Research, Roger-Gérard Schwartzenberg, had announced that he would more than double the 2001 TSE budget compared to its level of 2000.

In November 2000 the Minister also created the Groupement d'intérêt scientifique (GIS) for 'infections à prions' (Group of Scientific Interest into Prion Infection), which brings together the Ministry of Research, the Ministry of Agriculture, the Secretary of State for Health, public agencies such as the AFSSA and a number of research bodies. The remit of the GIS is to coordinate the action of the different stakeholders, to allocate TSE funding and to establish links with European research programmes.

The GIS has replaced the 1996 Expert Committee but is also headed by Dormont. It comprises an executive board (11 members who represent stakeholders) that sets the orientation of TSE research and allocates funding, and a scientific committee (15 members including at least 3 foreign researchers). Alperovitch, Hauw, Darlix, other biologists and a sociologist are members of the scientific committee, and so is James Ironside from the UK CJD Surveillance Unit.

In 2001, the number of TSE projects funded increased more than six-fold: 106 projects in 2001 compared to 16 in 2000. Forty nominated experts, including many more young researchers than in 1996, were responsible for the evaluation of research proposals. They came from a variety of organizations: INSERM, CNRS, INRA, CEA, AFSSA, the Pasteur Institute and different hospitals and universities.

The 106 funded projects fell into 7 thematic areas:

Theme 1: treatment of meat-and-bone meal (3 projects)
Theme 2: new therapeutic and diagnostic approaches (17 projects)
Theme 3: risk management (4 projects in social science)
Theme 4: human and animal epidemiology (5 projects)
Theme 5: pathogenesis (32 projects)
Theme 6: structural biology (11 projects)
Theme 7: cellular biology and nature of the agent (34 projects)

Studies on pathogenesis and cellular biology clearly outnumbered the others, and the focus on the nature of the agent was centre-stage. Nearly all projects on the nature of the agent included the term 'prion' in their title. Compared with the 1996 research programme, newcomers to the TSE field were now in a majority.

The situation that currently prevails in the French TSE field demonstrates that prion research has benefited immensely from a *political* event: the BSE crisis (see also Dressel, Chapter 3 this volume, for the German situation). Indeed, one may wonder whether the development of prion research would have been so spectacular without this unexpected event. These developments raise the question of whether the BSE crisis has also contributed to the settlement of the prion controversy. Does the over-representation of prion work in contemporary French TSE research indicate that the prion hypothesis is now widely accepted in France? In the following sections we will try to answer this question.

The prion controversy

We have seen that the first French TSE workers did not take up the prion hypothesis when Prusiner first formulated it in 1982. Even though his mentor had been involved in bringing to light phenomena that cast doubt on the viral nature of the TSE agent, Louis Court took up the non-conventional virus hypothesis. So did his student Dominique Dormont.

If the 1996 Expert Committtee and even more so the GIS have had a tremendously positive impact on prion research, Dormont's group continues to challenge the prion hypothesis. In 1997, his young colleague, Corinne Lasmézas, described experimental TSE in which PrP^{Sc} is barely or not detectable. She showed a dissociation between the accumulation of abnormal PrP, the neuropathological features and the appearance of clinical signs (Lasmézas et al., 1997). Scepticism appears useful in pointing out the shortcomings of the prion theory. However, during the same period, Prusiner was awarded the Nobel Prize for medicine and it appears to have been decisive in establishing the hegemony of the prion hypothesis (see also Dressel, Chapter 3 this volume).

During a symposium on 'The Truth in Science', held at the prestigious Collège de France on 16 October 2001, Dormont emphasized the difficulties encountered by researchers who want to publish results that do not follow the prion hypothesis. He argued that today the prion concept has become an obstacle to the growth of biological knowledge. Dormont is the best known French expert and is in a position of power due to his status as official expert adviser to the government, yet he seems unable to convince the scientific community that the nature of the TSE agent remains a mystery. Dormont is definitely a public figure in France; however, science is an international undertaking and most publications worldwide deal with the prion hypothesis. Consequently, newcomers to the field tend to endorse it to a much larger extent than the pioneers who have worked on the basis of the unconventional virus hypothesis. Even though a number of international scientists do not believe in the prion hypothesis, they do not express their reservations – at least publicly (see Dressel, Chapter 3 this volume).

Resistance to the prion hypothesis undoubtedly exists in France, especially in Dormont's and Darlix's laboratories. Nevertheless, the prion hypothesis is central in contemporary French research precisely because of the emergence of new actors. A number of young French scientists such as Sylvain Lehmann now collaborate with Prusiner's group and the prion hypothesis is increasingly accepted among them. For the younger generation, even though some questions remain as yet unresolved, this hypothesis is the most relevant approach. Some young scientists have even actively contributed to the enhancement of Prusiner's prestige in France. On 12 December 1996, Jean-Louis Laplanche invited Prusiner to Paris and succeeded in having him nominated doctor *honoris causa* at the Université Paris 5.

Several factors account for the widespread acceptance of the prion hypothesis, especially among new actors. As abundantly described in previous chapters, the viral nature of the scrapie agent was questioned as early as the 1960s, 15 years before Prusiner developed his prion theory (see Poulsen and Andersen, Chapter 1 this volume). Yet, in drawing the attention of the scientific community to a new entity – the prion – Prusiner succeeded in transforming an anomaly into a new way of conceiving a group of diseases. As shown by Mary Douglas, the classification of natural or cultural things is vital for whatever may be called 'structural' in social life: 'The labels [classifications] stabilize the flux of social life and even create to some extent the realities to which they apply ... ' (Douglas, 1986: 45).

As a result, despite disagreements, a common prion terminology has now gained currency in the TSE field (see Kim, Chapter 2 this volume). Crucially, this terminology is now being used even by opponents of the prion hypothesis such as Dormont (Dormont, 2002; Titeux et al., 2002).

Another reason why young researchers are easily drawn into prion territories is their lack of familiarity with the unconventional virus hypothesis, and especially their lack of contact with its most famous representatives, foremost among them Gajdusek. Gajdusek won the 1976 Nobel Prize for his work on kuru at a time when he firmly believed that TSEs were caused by a virus. We have seen that nearly all of the first French TSE workers visited his laboratory and there can be no doubt that his influence on them was considerable. Young scientists have not had this experience of working in a virus-oriented laboratory. Even Laplanche, who was introduced to TSEs at the end of the 1980s and is one of the 'oldest young' researchers, never had any contact with Gajdusek's group. He was attracted to the prion hypothesis from the start.[28]

It is well known that senior researchers in Edinburgh and elsewhere had unfortunate experiences with Prusiner (Rhodes, 1997), and this also happened to some French workers. In the early 1990s, Jean-Pierre Liautard, an INSERM researcher, sent a letter to Prusiner to inform him of his chaperone theory.[29] Though Prusiner requested further details he never referred to Liautard in his publications.[30] In contrast, the younger generation has not experienced the same problems in dealing with Prusiner. Soon after starting his TSE work, Laplanche began collaborating with David Westaway in Prusiner's laboratory. Laplanche realized that collaboration with Prusiner's group was easy. When he needed antibodies, he could obtain them without reciprocal arrangements, though he admits to being only a modest contributor in the field and certainly not a rival.[31] So, for younger scientists Prusiner is the undisputed 'master of the TSE field'. One additional factor that motivates their theoretical choice is 'aesthetic quality', by which they mean that the prion hypothesis is elegant, attractive and allows them to explore new and challenging ways of conducting research.

Another factor that explains the breakthrough of the prion hypothesis is peer review. Over the years Prusiner and his allies gained control over the referee system of the field. Prion researchers began increasing in numbers and supplanting the others. As a result, scientists who do not work on the prion hypothesis are now regarded as suspect by a number of referees. As pointed out by Dormont and Darlix, they face increasing difficulty in having their work published. In turn, the low numbers of papers drawing upon viral or other non-prion approaches now lead newcomers to the field to endorse the prion hypothesis almost mechanically. This, combined with the prestige Prusiner has gained from having been awarded a Nobel Prize, reinforces the impression that the prion controversy has been settled for good. It is worth mentioning that resistance is also weakening because the first TSE workers are now retired or approaching the end of their career (see Dressel, Chapter 3 this volume, for the German example). They can no longer offset the influence of Prusiner's supporters in the referee system.

The study of scientific controversies has long shown that the boundaries of science are not given. Boundary work is the notion devised by sociologist of science Thomas Gieryn to account for the fact that different actors produce different definitions of what is scientific and what is not. We have seen that in the prion controversy both sides performed boundary work, that is, tried to enhance or downplay the scientificity of the prion hypothesis. In publicly claiming that the prion concept now hinders knowledge, Dormont was performing boundary work. So was the Nobel Committee in awarding Prusiner the Nobel Prize for medicine.

This being said, we know that the boundaries of science are not drawn exclusively within the scientific community. Latour was among the first to show that those who are really doing science are not all at the bench (Latour and Woolgar, 1979; Latour, 1987). The closure of scientific controversies often involves actors outside the scientific community. In his study of the cold fusion controversy, Gieryn (1999) has shown how the media were brought into the fact-adjudication process by chemists Pons and Fleischmann whose claim to have discovered cold fusion was challenged by physicists. In the remainder of this chapter we will see that in France the hegemony of the prion hypothesis also stems, to some extent, from the media's boundary work.[32]

The media's boundary work

The concept of an infectious protein seemingly clashed with the 'central dogma of molecular biology' (Keyes, 1999). By the early 1990s, GH contamination and the BSE epidemic led the French media to cover the debate on the nature of the TSE agent. Later, with the 1996 British announcement on vCJD, the media were looking for an explanation of this new disease that threatened public health. From that time on they played a significant role in establishing the hegemony of the prion hypothesis in France.

Press coverage before 1996

Throughout the 1990s, the controversy on the nature of the TSE agent figured in French newspapers and magazines. In May 1990, the famous newspaper *Le Monde* addressed 'the mystery of mad cows' and the nature of the unconventional transmissible agent (Nau, May 1990). The protein-only hypothesis was mentioned but the term 'prion' was not yet used. On 17 July 1990, the possibility of BSE transmission to humans was raised in *Le Monde*. The scientific correspondent Jean-Yves Nau (who also wrote the previous article), explained that the nature of the agent was still controversial but mentioned the prion hypothesis (Nau, July 1990).

In 1992, a number of newspapers and magazines (*Le Monde, Le Figaro, Libération, La Recherche* and so on) gave an account of the 1992 report. The controversy was described in detail and Dormont's ideas were dominant: the nature of the agent was an enigma. Only the popular science magazine *Science et Vie* presented Prusiner's hypothesis as the most relevant one – on 30 October (Dubrana, 1992). On 31 January 1994, *Le Figaro* explained that the most likely hypothesis was the prion hypothesis (*Le Figaro*, 1994). Other articles were not so enthusiastic about the prion. However, after the popular science magazine *Pour la Science* published a paper by Prusiner in March 1995 (Prusiner, 1995), the press began to depict the prion hypothesis as the accepted one. On 7 April 1995, Jean-Yves Nau presented the prion as the infectious agent of 'prion diseases' in *Le Monde* (Nau, April 1995). Interestingly, on 14 February 1995 Nau was still questioning the nature of the agent (Nau, February 1995).

Press coverage after 1996

On 22 March 1996, just after the British announcement on the discovery of vCJD, the newspaper *L'Humanité* published an interview with French TSE scientist Jean-Jacques Hauw, who claimed that the most accepted hypothesis was that an infectious protein was the cause of TSEs (*L'Humanité*, 1996). On 27 March, Jean-Yves Nau (*Le Monde)* emphasized that the debate on the nature of the infectious agent was not over, even though the latest scientific results supported the prion hypothesis (Nau, March 1996). On 2 April, the popular science magazine *Eureka* presented Prusiner as a scientist who could be right against all other scientists (*Eureka*, 1996). On 26 April, an article in *Le Monde* was published with the headline: 'The risk of transmission of the bovine prion to humans is becoming clear' (Nau, April 1996). The article started with the question; 'Could the prion, responsible for "mad cow" disease, be the cause of CJD in humans?' The 'prion' was clearly presented as the infectious agent of these diseases. In the 'Mad Cow' special issue of *Science et Vie* in May 1996, the prion hypothesis was presented as dominant (*Science et Vie*, 1996). On 14 June 1996, according to the newspaper *Le Parisien*, the prion was responsible for TSEs (Darriulat et al., 1996). On 22 August 1996, *Le Monde* reported on Kurt Wüthrich's work on the spatial configuration of the prion, and journalist Jean-Yves Nau emphasized that the prion can be an infectious agent when its spatial configuration is abnormal (Nau, August 1996).

Here it is worth comparing such positive reporting with the state of affairs in the French scientific community. In September 1996, the Expert Committee headed by Dominique Dormont produced a report to propose a TSE research programme. The report admitted that for the viral explanation to be confirmed a virus obviously had to be identified. However, the report also emphasized Manuelidis's work which showed that the most infectious tissues were not necessarily those containing the highest levels of prion protein (see Kim, Chapter 2 this volume). The report went on to argue that in order to show that the prion protein can be infectious, PrP^{C} had to be isolated (or produced by a genetic device), modified to acquire the PrP^{Sc} conformation and injected into animals to cause disease. The report stressed that this experiment had not yet succeeded. The discrepancy between the status of the prion hypothesis in the scientific community and its coverage in the media shows that the media's activity was not restricted to reporting developments in TSE research. On the contrary, in constantly suggesting the superiority of the prion theory over other explanations, the media performed decisive boundary work.

In December 1996, *Le Figaro* used the headline 'How prions infect the brain' (Bader, 1996) to present Prusiner's latest experiments published in *Science* (Telling et al., 1996). Transgenic mice expressing normal human PrP had been injected with extracts of TSE brains. The journalist claimed that 'this experiment shows that the normal prion can be transformed into a pathological form'. Again, it is interesting to compare such reporting with the original claim. In their conclusion the authors of the *Science* paper argued:

> Our results provide a plausible mechanism of explaining diversity in a pathogen that lacks nucleic acid; the biological properties of prion strains seem to be encrypted in the conformation of PrP^{Sc} (...) Indeed, the foregoing data violate the widely and long-held idea that amino acid sequences are the sole determinants of the tertiary structures of biologically active proteins. (Telling et al., 1996: 2082)

Scientists' cautious phrasing was lost in *Le Figaro*, where prions were described as proteins causing TSEs.

In January 1997, *The Lancet* published a paper by John Collinge, the UK ambassador of the prion hypothesis (Hill et al., 1997). *Le Monde* reported: 'The prion responsible for the new variant of Creutzfeldt–Jakob disease was found in patients' tonsils' (Nau, 11 January 1997). Soon after, Dormont's group published in *Science* a paper on BSE transmission to mice in the absence of detectable abnormal prion protein (Lasmézas et al., 1997). Presumably motivated by national considerations, the press then questioned the prion hypothesis again. On 18 January 1997, *Le Monde*'s headline was: 'Frenchmen disrupt theories on the origin of "prion diseases"' (Nau, 18 January 1997). On 13 March 1997, the weekly magazine *L'Express* reported that a French group was challenging the prion's role in CJD (Casteret, 1997). On 17 January and 17 July 1997, *Le Figaro* pointed out that experts had conflicting views on the mysterious infectious agent (Petitnicolas, January and July 1997). On 18 September 1997, *Le Monde* emphasized that the scientific controversy was not closed (Folléa, 1997). On 8 October 1997, the newspaper *La Croix* featured this headline: 'Prusiner's thesis provokes debate' and reported that not all scientists agreed with the prion hypothesis (Verdier, 1997).

If the Dormont group's 1997 *Science* paper impacted on the media's boundary work, it did not last very long. Prusiner was awarded his Nobel Prize in October 1997. On 8 October, *Le Monde* reported that the Nobel Prize for medicine had been granted to the 'discoverer of prions'. Prions were described as the agents known to cause TSEs (Nau, October 1997), though their possible association with genetic material was mentioned. Thereafter, the hegemony of the prion hypothesis was established in the French media and was never challenged again. Dormont's advisory position did not offset the prestige and authority Prusiner gained from his Nobel Prize.

For the French media Prusiner's theory became the orthodoxy. The best illustration lies in the events surrounding the symposium on TSEs organized in Paris by the French Academy of Sciences on 14–16 March 2001. The world's most famous TSE experts attended it. During the symposium Kurt Wüthrich from Zurich put forward: 'We don't know if PrP is the infectious agent. It just deposits in diseased brain' (Wüthrich, verbal communication). None of the scientists present objected to this claim, not even Prusiner. Yet, on 22 March 2001 *Le Monde* published an interview with Prusiner who declared: 'Today it is known that the causal agent of infections is the prion protein' (*Le Monde*, 2001). The article made no reference whatsoever either to Wüthrich or to Dormont. Both had disappeared behind a Nobel laureate.

The interview in *Le Monde* is a good opportunity to look at Prusiner's contextual strategy in the prion controversy. Sociologists of science such as Latour and Woolgar (1979) and Gieryn and Figert (1990) have pointed out the interpretative flexibility of science and the contextual nature of its boundaries. Equipped with the prestige associated with his

Nobel Prize, Prusiner enrolled a newspaper to claim that the prion protein was the causal agent of TSEs. However, his boundary work was carefully adjusted to local conditions. In a paper published in *Nature* in the same year, his claim was toned down: '*Prions* are the transmissible pathogenic agents responsible for diseases such as scrapie and bovine spongiform encephalopathy' (Peretz et al., 2001: 739).

Here Prusiner was using the term 'prions' in its generic meaning of infectious particles that resist treatment that destroy nucleic acid, rather than in the specific meaning of an infectious protein. This was nothing new. Prusiner has always been aware of the importance of local conditions and has always cautiously worded the claims he makes in scientific journals. Years ago he told Gary Taubes: 'I never said it's only an infectious protein. I've never said that in one paper. You'll not find it. I've been very, very careful' (Taubes, 1986: 52).

In his analysis of the media's boundary work Gieryn has noted:

> When the mass media take upon themselves the task of distinguishing genuine scientific knowledge from putatively less responsible claims scientists whose claims were made suspect will redraw the cultural map to restore a monopoly over such cartographic efforts to those inside science. (Gieryn, 1999: 17)

This happened in the cold fusion controversy. Journalists were eventually excluded from the fact-adjudication process by physicists seeking to denounce Pons and Fleischmann's claim to a discovery. Nothing similar was observed in relation to the prion controversy and the media's boundary work in France. Scientists who are sceptical of the prion hypothesis, especially Dormont, did not try to restore the monopoly of the TSE community, not even after Prusiner's interview in *Le Monde*.

Since the British announcement that established a causal link between BSE and vCJD, the French media have been looking for an explanation of these dreadful and strange diseases threatening animal and human health. Prusiner and his prions provided them with the answer.

Conclusion

For a long time TSE research in France struggled along and depended on individual initiatives. TSEs were rare diseases and French authorities did not find it necessary to establish a proper research structure. In the 1960s, in collaboration with Tikvah Alper, the French radiobiologist

Raymond Latarjet brought to light the extreme resistance of the scrapie agent and questioned its viral nature. Though the AHS was interested in the resistance of the scrapie agent to radiation, basic TSE research was not prioritized. In the 1970s and 1980s, a few researchers continued to study the nature of the agent but did not take up the protein-only hypothesis. By 1992, only a handful of research centres were involved in the study of TSEs and the prion hypothesis was not predominant.

This situation changed dramatically from 1996. French TSE research was inadvertently stimulated by an external, political, event – the BSE/vCJD crisis. Public concerns forced the French authorities to address the problems raised by these diseases. A TSE research programme was then set up and research funds were released. Several scientists, including many newcomers to the field, took advantage of the availability of TSE funding. This immediately invigorated the development of prion research in France. If prion work contributed to placing the prion hypothesis one length ahead of the viral model, national factors contributed to keeping a virus school alive and the scientific community was divided with respect to the nature of the infectious agent.

However, the second and massive round of funding of 2001 has now resulted in the hegemony of the prion hypothesis in France. A number of new and young researchers have invaded the promising TSE field. Their lack of familiarity with the viral model, the low exposure of this model and alternative explanations in scientific literature and in peer review, and Prusiner's Nobel Prize are factors that explain why they enthusiastically base their work on the prion hypothesis.

Acceptance that an infectious protein is the agent of TSEs has also been fuelled by another external factor – the boundary work performed by the French media. Throughout the 1990s, the press reported on GH contamination and on the BSE crisis. With the discovery of vCJD in 1996, it became imperative for the media to offer an explanation of these diseases. Since media coverage is shaped by structural relationships within communities, Prusiner's dominant position, especially after 1997, has been echoed by a change in reporting. The press started to participate actively in establishing the hegemony of the prion theory, and the controversy sustained by the lack of final confirmation faded away. For the French media the TSE agent is now the prion protein.

A few years ago the French press portrayed Stanley Prusiner as 'a genius who proposed, alone against all, a revolutionary concept, a complete new way to approach the question of the causal agent of TSEs, and who eventually succeeded in convincing other scientists' (*Eureka*, 1996). It is to be hoped that in highlighting the role of the public sphere in the success of prion research and the prion hypothesis in France, the present chapter has put into perspective the mythical story of this lone genius.

Acknowledgements

I would like to thank INSERM (Institut National de la Santé et de la Recherche Médicale) for financial support, and to express my gratitude to Eve Seguin and Yoheved Berwald-Netter for comments, rereading and English corrections.

Notes

1. It was not the case during the eighteenth century. Wool was one of the most important products of the English economy and in 1755 a debate on the economic consequences of scrapie and the need for government intervention took place in the British Parliament (Brown and Bradley, 1998: 1688).
2. The AHS is comprised of medical services, hospitals and research centres.
3. Interview with Dominique Dormont, 30 December 1997, Fontenay aux Roses.
4. Interviews with Françoise Cathala, 9 January and 7 April 1998, Paris; interview with Louis Court, 22 June 1998, Paris.
5. Ibid.
6. He was then a young lieutenant and an army medical officer.
7. Interview with Dominique Dormont, op. cit.
8. DRET contract no. 88–1030 conducted by Louis Court, *Epidémiologie et biologie d'affections virales à évolution lente du système nerveux central*, final draft from L. Court, D. Dormont, J. L. Deslys, October 1988, AHS archives 552 [76].
9. For example, Dormont et al., *Inactivation du virus de l'immunodéficience humaine et des virus lents non conventionnels dans les produits biologiques d'origine humaine utilisés en thérapeutique*, SSA (Service de santé des armées) 1988, Trav. Scient. no. 9, AHS archives, 552 [25].
10. DRET was part of DMA (Délégation Ministérielle pour l'Armement).
11. Interview with Louis Court, op. cit.
12. The 1992 report, op. cit.: 95.
13. Ibid.: 5
14. Ibid.: 38–61.
15. Laplanche had links with David Westaway from Prusiner's group and with John Collinge, the ambassador of the prion hypothesis in the UK.
16. The 1992 report, op. cit.: 40.
17. Ibid.: 51.
18. Ibid.: 52.
19. Ibid.: 53.
20. Ibid.: 55.
21. Ibid.: 58.
22. Ibid.: 60–1.
23. Ibid.: 72.
24. Ibid.: 77.
25. Ibid.: 117.

26. Telephone interview with Jean-Luc Darlix, 4 February 2003.
27. As we will see below with Lasmézas and colleagues' 1997 paper.
28. Interview with Jean-Louis Laplanche, 5 July 2001, Paris.
29. Liautard proposed a model based on theoretical considerations involving molecular chaperones, namely proteins involved in the folding of proteins. He explained that prions could be molecular chaperones that are required for their own assembly (auto-chaperones) (Liautard, 1991). This theory derived from an analysis of protein folding and its consequences were explored by computer simulation. Thermokinetic analysis of protein folding showed that a misfolded chaperone gave rise to new misfolded chaperones.
30. Interview with Jean-Pierre Liautard, 9 May 2001, Strasbourg.
31. Interview with Jean-Louis Laplanche, op. cit.
32. As suggested by Taubes (1986), this might apply to the USA as well.

References

Aucouturier, P. et al. (2001) 'Infected Splenic Dendritic Cells are Sufficient for Prion Transmission to the CNS in Mouse Scrapie', *J. Clin. Invest.* 108 (5): 703–8.

Barrairon, E. (1990) 'J. Cuillé et P. L. Chelle "inventent" dans les années trente la notion de maladies virales à évolution lente', *Revue Méd. Vét.* 141: 575–9.

Beauvais, P. and T. Billette de Villemeur (1996) *Maladie de Creutzfeldt–Jakob et autres maladies à prion*. Paris: Médecine-Sciences, Flammarion.

Brizay, A. et al. (1997) *La 'vache folle': d'une crise sanitaire à une crise politique et économique*. Paris: ENGREF (Ecole Nationale du Génie Rural des Eaux et Forêts).

Brown, P. and R. Bradley (1998) '1755 and All That: a Historical Primer of Transmissible Spongiform Encephalopathy', *British Medical Journal* 317: 1688–92.

Brown, P. et al. (1987) 'The Epidemiology of Creutzfeldt-Jakob Disease: Conclusion of a 15-year Investigation in France and Review of the World Literature', *Neurology* 37: 895–904.

Cathala, F. et al. (1985) 'Failure to Detect Scrapie Virus in Sheep at Slaughter in a Highly Endemic Region of France', *Eur. J. Epidemiol.* 1: 90–3.

Chabry, J. et al. (1998) 'Specific Inhibition of in vitro Formation of Protease-resistant Prion Protein by Synthetic Peptides', *J. Biol. Chem.* 273 (21): 13203–7.

Chabry, J. et al. (1999) 'Species-independent Inhibition of Abnormal Prion Protein (PrP) Formation by a Peptide Containing a Conserved PrP Sequence', *J. Virol.* 73 (8): 6245–50.

Chamak, B. (1999) 'La vache folle: une crise annoncée', *Pour la Science* 256: 12–16.

Clément, J. M. et al. (1992) *Rapport sur l'hormone de croissance et la maladie de Creutzfeldt–Jakob*. Inspection Générale des Affaires Sociales, SA 19, no. 92145, December.

Collinge, J. (2001) 'Prion Diseases of Humans and Animals: Their Causes and Molecular Basis', *Annu. Rev. Neurosci.* 24: 519–50.

Court, L. (ed.) (1983) *Virus non conventionnels et affections du système nerveux central*. Paris: Masson.

Court, L. (1996) 'L'étude des encéphalopathies spongiformes (ou maladies à prion) au service de santé des armées', *Lyon Val* 58: 12–17.

Cuillé, J. and P. L. Chelle (1936) 'La maladie dite tremblante du mouton est-elle inoculable?', *Compte Rendu Acad. Sci.* 203: 1552–4.

Dormont, D. (1992) *Les encéphalopathies subaiguës spongiformes humaines et animales: description clinique et biologique, facteurs étiologiques, conséquences sur la santé publique, et axes de recherches développés en France.* Rapport au Ministre de la recherche et de l'espace, 30 June.

Dormont, D. (2002) 'Prion Diseases: Pathogenesis and Public Health Concerns', *FEBS Lett.* 529 (1): 17–21.

Douglas, M. (1986) *How Institutions Think.* Syracuse: Syracuse University Press.

Dron, M. et al. (2000) 'Mouse Scrapie Responsive Gene 1 (Scrg1): Genomic Organization, Physical Linkage to sap30, Genetic Mapping on Chromosome 8, and Expression in Neuronal Primary Cell Cultures', *Genomics* 15 (70): 140–9.

Gabus, C. et al. (2001a) 'The Prion Protein has DNA Strand Transfer Properties Similar to Retroviral Nucleocapsid Protein', *Mol. Biol.* 307 (4): 1011–21.

Gabus, C. et al. (2001b) 'The Prion Protein has RNA Binding and Chaperoning Properties Characteristic of Nucleocapsid Protein NCP7 of HIV-1', *J. Biol. Chem.* 276 (22): 19301–9.

Gajdusek, C. and V. Zigas (1957) 'Degenerative Disease of the Central Nervous System in New Guinea', *New Engl. J. Med.* 257: 974–8.

Gieryn, T. (1999) *Cultural Boundaries of Science – Credibility on the Line.* Chicago: University of Chicago Press.

Gieryn, T. and A. Figert (1990) 'Ingredients for a Theory of Science in Society: o-rings, Ice Water, c-clamp, Richard Feynman and the Press', in S. Cozzens and T. Gieryn (eds) *Theories of Science in Society.* Bloomington: Indiana University Press: 67–97.

Hill, A. et al. (1997) 'Diagnosis of New Variant Creutzfeldt–Jakob Disease by Tonsil Biopsy', *Lancet* 349 (9045): 99–100.

Hirsch, M. et al. (1996) *L'affolante histoire de la vache folle.* Paris: Balland.

Hornsey, S. and J. Denekamp (1997) 'Tikvah Alper: an Indomitable Spirit', *Int. J. Radiat. Biol.* 71 (6): 638.

Hsiao, K. et al. (1989) 'Linkage of a Prion Protein Missense Variant to GSS', *Nature* 338: 342–5.

Kellershohn, N. and M. Laurent (2001) 'Prion Diseases: Dynamics of the Infection and Properties of the Bistable Transition', *Biophys. J.* 81 (5): 2517–29.

Keyes, M. (1999) 'The Prion Challenge to the "Central Dogma" of Molecular Biology, 1965–1991', *Stud. Hist. Biol. & Biomed. Sci.* 30 (1): 1–19.

Komar, A. et al. (1997) 'Differential Resistance to Proteinase K Digestion of the Yeast Prion-like (Ure2p) Protein Synthesized in vitro in Wheat Germ Extract and Rabbit Reticulocyte Lysate Cell-free Translation Systems', *FEBS Lett.* 415 (1): 6–10.

Komar, A. et al. (1999) 'The [URE3] Yeast Prion: from Genetics to Biochemistry', *Biochemistry* 64 (12): 1401–7.

Lasmézas, C. et al. (1997) 'Transmission of the BSE Agent to Mice in the Absence of Detectable Abnormal Prion Protein', *Science* 275: 402–5.

Latarjet, R. et al. (1970) 'Inactivation of the Scrapie Agent by Near Monochromatic Ultraviolet Light', *Nature* 227 (5265): 1341–3.

Latour, B. (1987) *Science in Action.* Cambridge: Harvard University Press.

Latour, B. and S. Woolgar (1979) *Laboratory Life: the Social Construction of Scientific Facts.* Los Angeles/London: Sage Publications.

Liautard, J. P. (1991) 'Are Prions Misfolded Molecular Chaperones?', *FEBS Lett.* 294 (3): 155–7.

Mattéi, F. (dir.) (1997) *De la 'vache folle' à la 'vache émissaire'*. Rapport Assemblée Nationale, no. 3291.

Oesch, B. et al. (1985) 'A Cellular Gene Encodes Scrapie PrP 27-30 Protein', *Cell* 40: 735–46.

Peretz, D. et al. (2001) 'Antibodies Inhibit Prion Propagation and Clear Cell Cultures of Prion Infectivity', *Nature* 412: 739–43.

Petitjean, S. (1996) *La crise de la 'vache folle'*. Les dossiers de l'environnement de l'INRA: vache folle, no. 13.

Prusiner, S. (1995) 'Les maladies à prions', *Pour la Science* 209: 42–50.

Rangon, C. M. et al. (2003) 'Different Chromatogranin Immunoreactivity between Prion and A-beta Amyloid Plaque', *Neuroreport* 14 (5): 755–8.

Rhodes, R. (1997) *Deadly Feasts*. New York: Simon & Schuster.

Schwartz, M. (2001) *Comment les vaches sont devenues folles*. Paris: Odile Jacob.

Taubes, G. (1986) 'The Game of the Name is Fame. But Is it Science?', *Discover* December: 28–52.

Telling, G. et al. (1995) 'Prion Propagation in Mice Expressing Human and Chimeric PrP Transgenes Implicates the Interaction of Cellular PrP with Another Protein', *Cell* 83: 79–90.

Telling, G. et al. (1996) 'Evidence for the Conformation of the Pathological Isoform of the Prion Protein Enciphering and Propagating Prion Diversity', *Science* 274: 2079–82.

Thual, C. et al. (1999) 'Structural Characterization of Saccharomyces cerevisiae', *J. Biol. Chem.* 274 (19): 13666–74.

Thual, C. et al. (2001) 'Stability, Folding, Dimerization, and Assembly Properties of the Yeast Prion Ure2p', *Biochemistry* 40 (6): 1764–73.

Titeux, M. et al. (2002) 'Differences in the Activation of the GFAP Gene Promoter by Prion and Viral Infections', *Brain Res. Mol. Brain Res.* 109 (1–2): 119–27.

Vilette, D. et al. (2001) 'Ex vivo Propagation of Infectious Sheep Scrapie Agent in Heterologous Epithelial Cells Expressing Ovine Prion Protein', *Proc. Natl. Acad. Sci.* 98 (7): 4055–9.

Weissmann, C. (1991) 'A "Unified Theory" of Prion Propagation', *Nature* 352: 679–83.

Press review

Bader, J. M. 'Comment les prions infectent le cerveau', *Le Figaro*, 26 December 1996: 10.

Casteret, A. M. 'Creutzfeldt–Jacob: et si c'était un virus ... ', *L'Express*, 13 March 1997: 25.

Darriulat, J. et al. 'L'état des connaissances', *Le Parisien*, 14 June 1996: 3.

Dubrana, D. 'Le syndrome de la vache folle', *Science et Vie*, 30 October 1992: 52–7.

Eureka 'Vache folle, maladie de Creutzfeldt–Jakob: le dossier scientifique', 2 April 1996: 31.

Folléa, L. 'L'hypothèse de la transmission à l'homme de la maladie de la "vache folle" se précise', *Le Monde*, 18 July 1997: 16.

Le Figaro 'Ce prion inconnu qui fait du cerveau une éponge', 31 January 1994: 15.

L'Humanité 'L'avis d'un expert: il n'y a pas de risque épidémique', 22 March 1996: 8.

Le Monde 'Il est trop tôt pour avoir ne serait-ce qu'une idée de l'évolution de l'épidémie', Prusiner's interview with Paul Benkimoun and Elisabeth Burseaux, 22 March 2001: 8.

Nau, J. Y. 'Le mystère de la vache folle', *Le Monde,* 30 May 1990: 22.
Nau, J. Y. 'Après la publication d'une étude dans "The Lancet" une controverse se développe sur les risques de transmission à l'homme de la maladie des "vaches folles" ', *Le Monde,* 17 July 1990: 14.
Nau, J. Y. 'Deux cas de maladie de Creutzfeldt–Jakob inquiètent les responsables sanitaires', *Le Monde,* 14 February 1995: 15.
Nau, J. Y. 'Les mortels mystères des maladies à prions', *Le Monde,* 7 April 1995: 24.
Nau, J. Y. 'Le dérangeant mystère des maladies à prions', *Le Monde,* 27 March 1996: 3.
Nau, J. Y. 'Les risques de transmission du prion bovin à l'homme se précisent', *Le Monde,* 26 April 1996: 21.
Nau, J. Y. 'La maladie de la "vache folle" dans la troisième dimension des prions? ', *Le Monde,* 22 August 1996: 13.
Nau, J. Y. 'Le prion responsable de la nouvelle forme de maladie de Creutzfeldt–Jakob a été retrouvé dans les amygdales des victimes', *Le Monde,* 11 January 1997: 25.
Nau, J. Y. 'Des Français bousculent les théories sur l'origine des "maladies à prions" ', *Le Monde,* 18 January 1997: 25.
Nau, J. Y. 'Le prix Nobel de médecine est attribué au découvreur des prions', *Le Monde,* 8 October 1997: 27.
Petitnicolas, C. 'Vache folle: la théorie du prion malmenée', *Le Figaro,* 17 January 1997: 12.
Petitnicolas, C. 'De la vache folle à l'homme: une transmission limitée', *Le Figaro,* 17 July 1997: 9.
Science et Vie 'Vaches folles: l'intox', 944, May 1996: 100.
Verdier, M. 'La thèse de Prusiner suscite des débats', *La Croix,* 8 October 1997: 10.

6
The UK BSE Crisis and the Prion Discursive Chain in Scientific Literature

Eve Seguin

Prion discourse and the BSE crisis in the UK

For decades transmissible spongiform encephalopathies (TSEs) were obscure diseases that attracted the attention of a handful of scientists worldwide. Apart from sheep scrapie which has been endemic in UK herds for over two centuries, their incidence is extremely low. Kuru affected only the allegedly cannibalistic Fore tribe in New Guinea and is now disappearing.[1] Sporadic CJD affects only one person in a million worldwide, mainly in the age category 60 and over.

TSEs, their elusive agent and their victims could have remained for ever a relatively private matter occupying just a few scientists. Yet, the discovery in November 1986 of bovine spongiform encephalopathy (BSE), a new TSE affecting UK cattle herds, changed this situation. The BSE epidemic and its potential risk for human health turned TSEs into a major public concern and stimulated the development of TSE research. As shown by Chamak (Chapter 5 this volume) and Dressel (Chapter 3 this volume), there is no doubt that BSE contributed to the expansion of prion work and offered prion discourse a unique opportunity for spreading into the public domain.

In turn this infiltration of prion discourse into public health raises the question of the political effects it had on the management of the BSE epidemic. If from 1982 TSE research underwent a remarkable transformation, few scholars, if any, have endeavoured to put it in relation with the equally dramatic transformation of the risk assessment of BSE. For years, UK scientists and officials, both at the Ministry of Agriculture and Department of Health, believed that humans were not susceptible to BSE (Phillips Report, 2000).[2] This belief was shared by the three independent scientific advisory committees on TSEs successively set up by the UK government: the Southwood Committee (1988), the Tyrrell Committee (1989) and the Spongiform Encephalopathies Advisory Committee (SEAC) (April 1990 onwards). For years the official line was 'British beef is safe'.

In May 1990 a domestic cat was diagnosed with feline spongiform encephalopathy. This case caused public anxiety and was subjected to intense media coverage. Nevertheless, it did not change the reassurance strategy. The Chief Medical Officer and the Ministry of Agriculture issued press releases to allay public fears. The Chief Medical Officer and the Chief Veterinary Officer told the media that there was no cause for concern. Minister of Agriculture John Gummer fed his daughter a beefburger on television (Phillips Report, 2000). At the Chief Medical Officer's request SEAC produced a 12-page opinion which confirmed that it was safe to eat British beef (SEAC, 1990). Even though TSEs were later discovered in many species not previously affected, SEAC's risk assessment of BSE remained the same. For years the social discourse of the TSE field did not allow scientists to claim that BSE posed a risk to human health without discrediting themselves (Seguin, 2002).

Suddenly, in March 1996 SEAC declared that the ten cases of a new variant of CJD (vCJD) discovered in the UK were caused by the consumption of BSE-infected beef. By their own admission they had no scientific evidence to support this claim (SEAC, 1996). The accuracy and timeliness of their statement were open to discussion and were questioned by several scientists (Seguin, 2000). In particular, it was not known if the two diseases were caused by the same agent.[3] Thus, it appears that the discovery of vCJD in itself does not account for the March 1996 statement, which is better construed as a *discursive event* (Seguin, 2004).

Prion discourse may have played a role in this event. From its inception it dramatically differed from conventional TSE discourse, and this was not limited to terminology. Prion discourse draws upon several linguistic and textual features to convey an original construction of TSEs and their science. One such feature – molecular experimental language – suggests that a 'simple' answer can be provided to any question. Following a recommendation of the Southwood Report (Working Party on BSE, 1989),[4] in December 1989 neurologist Robert Will submitted to the Department of Health a grant proposal for the setting up of a CJD Surveillance Unit, which was successful. The very long incubation period of TSEs was a central element of TSE discourse and the grant proposal stated that to exclude BSE transmission to humans *epidemiological* monitoring of the population would be necessary for 15–20 years (Will, 1989). Prion discourse suggested a much more economic solution: 'Because our studies with *transgenic* mice show that the species barrier for passage of scrapie prions between rodent species is likely to reside in the amino-acid sequence of PrP, similar *experiments* seem relevant in examining the potential for transmission of prions from beef and sheep products to humans' (Westaway and Prusiner, 1990: 113; my italics). The contrast between the two discourses is stark.

In an outstanding study Faye (1972) has challenged conventional explanations of the rise of Nazism, and shown that the intense circulation of extreme-right discourses in the Weimar republic made the unacceptable acceptable: it legitimized Nazism and its horrendous policies. Faye's insistence on discursive circulation is relevant for understanding the political function of science (Seguin, 2001), and the BSE crisis in particular. We may hypothesize that the circulation of the prion construction of TSEs changed the perception of BSE and made its transmissibility to humans politically acceptable. In March 1996 it eventually allowed SEAC to take the risk of claiming, despite the lack of scientific evidence, that vCJD represented BSE transmission to humans (Seguin, 2004).

The circulation of prion discourse can be documented since it materialized in 'texts'[5] and formed a discursive chain. The prion chain primarily comprised the discursive production of Prusiner and prion researchers. However, due to interdiscursivity, the importance of which has been demonstrated by discourse analysis, the prion chain also included texts that were not produced by prion researchers. Indeed, in the 1980s most users of prion terminology were not prion supporters but scientists who qualified it and used alternative terms. We may therefore define the prion chain as being made up of all those texts that bore *traces* of prion discourse, from the simple mention of Prusiner's name to the full prion construction of TSEs.

During the years up to March 1996 SEAC were potentially exposed to the prion chain via several channels of communication: regular review of the BSE R&D programme set up by the UK Ministry of Agriculture; presentations made by various scientists at SEAC meetings; discussion with scientists and representatives of UK research councils and other organizations; correspondence with various actors; scientific collaboration of some SEAC members with prion supporters, etc. Another important channel was formal publication since part of SEAC's remit was to review TSE literature. At SEAC meetings key articles and updated lists of papers were tabled for discussion. Individual members also carried out private searches. SEAC and one of its predecessors, the Tyrrell Committee, published reports on TSE research in 1989, 1992 and 1995, which necessitated the collaboration of external scientists for input of additional references (Pattison, 1998; SEAC, 1998).

BSE, Prusiner and scientific literature

An important episode demonstrates that from the very beginning of the BSE saga, scientists involved in the management of the epidemic were confronted with the prion chain in scientific literature. The discovery of BSE was reported to the scientific community in an article published by Gerald Wells, head of the neuropathology section at the Central Veterinary Laboratory, the scientific arm of the UK Ministry of Agriculture (Wells et al., 1987). From that time on Wells was a key player on the governmental BSE R&D programme. Two months after the publication, Prusiner sent Wells a devastating criticism of the article, and added:

> We were also surprised at the list of references. It seems to us that you have certainly overlooked a large body of the scrapie literature; perhaps you are unaware of this work. Assuming that is the case, I have enclosed a couple of reviews: one that was published in *Lab Investigation* by Steve DeArmond and myself last year, as well as one which will be published next week in the *New England Journal of Medicine*. (Prusiner, 1987: 1)

Prusiner did not content himself with sending his own publications but also one that cited him: 'I have also enclosed for your reading a recent scholarly review by Ted Diener' (Prusiner, 1987: 1).

Instead of openly blaming Wells for not taking his work into account, Prusiner drew upon the contingent repertoire (Gilbert and Mulkay, 1984) and blamed him for being an incompetent scientist. He used scientific literature as a weapon and even physically mobilized it. His letter ensured that the discourse on BSE could not dispense with prion literature. Wells felt obliged to reply: ... I think perhaps I may have read a little more than you surmise from the reference list in the paper (Wells, 1988: 1).

This episode raises several questions. To what extent and where did prion discourse circulate in scientific communities? What was the progression of the prion chain in the UK up to the March 1996 announcement? What kind of exposure to the prion chain did SEAC experience while reviewing TSE literature? To give preliminary answers to these questions the remainder of this chapter is devoted to a quantitative study of the prion discursive chain in scientific literature. Using data provided by the Web of Science,[6] we will analyse Prusiner's publications and citations between 1982 (when he coined the term 'prion') and 1995. Before proceeding to this study, however, a brief review of the uses and shortcomings of bibliometrics is necessary.

Bibliometrics

In the 1960s, sociologists of science started to assess the scientific contribution of scientists with two indicators: their productivity (publications) and its quality and influence (citations).[7] They regarded the numbers of publications and citations as objective measurement of the progress of science. Since then bibliometrics has had a strange development. It has become a field of specialization in its own right, however, it is not accepted by the research community itself (Godin, 2002). In science and technology studies it is now deeply unfashionable since its main assumptions have been severely criticized.[8]

On the one hand, the measurement of productivity is based on the assumption that formal publication is the principal means used by scientists to communicate their work to others (Frank Fox, 1983). Yet, it has been shown that many crucial events that contribute to shaping the development of scientific fields are due to informal communications (Edge, 1979). One striking example is the knowledge of how to do things, be it experimental design or technological artefacts. This has been called tacit knowledge because far from being described in scientific papers, it remains invisible and inarticulate (Collins, 1974). Other scholars question the importance of publication in science since nowadays the vast majority of scientific papers go unread (Fuller, 2000).

On the other hand, the measurement of influence is based on the assumption that scientists acknowledge their debts via formal citation, and that those that are cited most by others have had the most impact (Garfield, 1970). However, it has been shown that citations vary in function. For instance, a significant proportion of citations are perfunctory, that is, they simply indicate that some work has been performed in the same area. This observation has led authors to propose various citation typologies (Moravcsik and Murugesan, 1975; Chubin and Moitra, 1975). It follows that taking any citation as an indicator of influence is erroneous. Besides, different kinds of discrepancies exist between influence captured by the bibliography and influence indicated in the text. For instance, several sources of influence can be concealed by the citation of a single review paper. Finally, and most importantly, citation concerns only the formal level of scientific communication. When the informal level is taken into account, it has been estimated that only about 15 per cent of the influence exerted on a paper is captured by citation counts (MacRoberts and MacRoberts, 1986).

Quantitative analysis of the prion chain in scientific literature

The above arguments are undoubtedly founded in their critique of bibliometrics as a self-proclaimed objective and superior way of studying scientific developments. However, they do not invalidate bibliometric methods as such, as one of the most vocal critics has admitted (Edge, 1979). The relevance of this approach depends on the uses it is put to. As with any research method, the purpose of the exercise should always be theoretically defined. In the present study it should be absolutely clear that bibliometrics is not aimed at assessing Prusiner's scientific 'contribution' or 'influence' but at mapping the prion chain in scientific literature.

Prusiner's publications

Prusiner's publications are indicators of the circulation of prion discourse in scientific literature in two different ways. On the one hand, they signal that prion discourse got out of Prusiner's laboratory and personal networks, and constitute evidence of exposure of editorial boards and referees of the journals where his articles were published. On the other hand, they indicate exposure of readers of these journals. It is well known that nowadays most scientific articles are not read (Fuller, 2000), yet a proviso must be made here. Titles of Prusiner's publications themselves make abundant use of prion terminology and may therefore contain features that contribute to the prion construction of TSEs. It follows that the readership of these journals may well have been exposed to prion discourse even if Prusiner's papers were not actually read.

Consultation of the Web of Science reveals that between 1982 and 1995 Prusiner published 319 papers.[9] This figure derives from the normal count procedure, which allocates credit to all contributors of a multi-authored publication. In other words, Prusiner is regarded as the author of all papers that bear his name, including the numerous ones he co-authored. Critics have argued that the normal count procedure is not a fair measure of production because it inflates the score of those who produce multi-authored publications (Lindsey, 1980). This

Table 6.1 Prusiner as first and sole author of his publications, 1982–95

Year	*No. papers*	*No. first*	*% first*	*No. sole*	*% sole*
1982	14	6	42.8	2	14.2
1983	12	3	25.0	0	0
1984	13	4	30.7	3	23.0
1985	17	3	17.6	1	5.8
1986	33	1	3.0	3	9.0
1987	32	2	6.2	4	12.5
1988	21	2	9.5	1	4.7
1989	13	0	0	1	7.6
1990	32	1	3.1	3	9.3
1991	23	2	8.6	2	8.6
1992	21	0	0	2	9.5
1993	25	3	12.0	8	32.0
1994	32	2	6.2	7	21.8
1995	31	4	12.9	3	9.6
All years	319	33	10.3	40	12.5

observation is of limited relevance here since we are not trying to determine Prusiner's production but the number of core units that made up the prion chain. It should also be noted that using the normal count does not amount to suggesting that Prusiner has written most of the 319 articles. Contrary to a commonsensical assumption, high-profile scientists are seldom first author of their multi-authored publications. This is confirmed by Table 6.1 which shows that between 1982 and 1995 Prusiner was first or sole author of only 22.8 per cent of his papers.

The publication of 319 papers in 14 years seems to be an outstanding achievement; however, the only way to confirm this impression is to compare Prusiner to other TSE scientists in the same period. In this comparison the normal count is also used for other scientists, that is, they too are granted the full value of their multi-authored papers.[10] Table 6.2 shows that Prusiner's number of publications is superseded only by Carleton Gajdusek's, who also won a Nobel Prize in 1976 for his work on kuru. The only other scientist whose number approaches Prusiner's is Joe Gibbs, who was Gajdusek's close collaborator. All other TSE researchers are a far cry behind.

This comparison clearly shows that 319 papers in 14 years is a very high number indeed, and suggests that the prion chain was extensive, at least in the form of core units.

Table 6.2 Numbers of papers by TSE researchers, 1982–95

A. G. Dickinson (retired 1987)	18
M. E. Bruce	29
P. A. Merz	34
B. Caughey	43
R. H. Kimberlin	46
R. I. Carp	62
L. Manuelidis	76
R. M. Ridley	78
B. Chesebro	82
C. J. Gibbs	239
S. B. Prusiner	319
D. C. Gajdusek	349

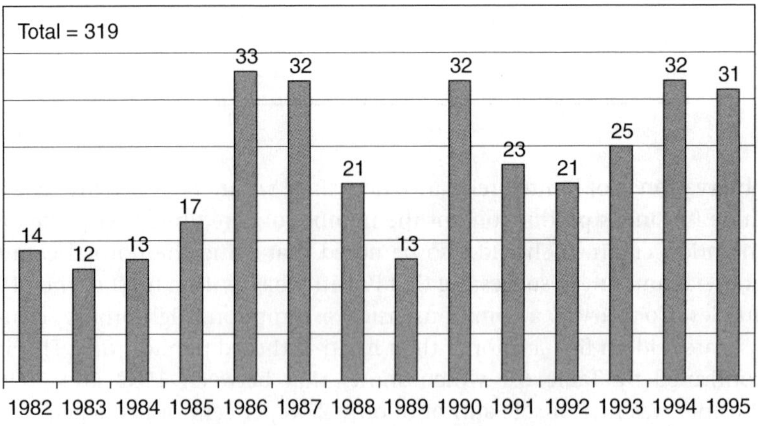

Figure 6.1 Annual numbers of papers published by Prusiner, 1982–95

The distribution of these 319 papers in time will give us a more detailed picture. The perspective of cumulative advantage documented by Merton and others holds that scientists who experience early success are able to obtain the increased time, facilities and support for continued research (Frank Fox, 1983). From this perspective we could expect Prusiner to have published an increasing number of papers in the period under scrutiny. Yet, Figure 6.1 shows that his production did not follow a linear pattern. The numbers follow a random progression with three peaks: 1986–87, 1990, 1994–95. This is an interesting finding: the circulation of prion discourse was as widespread in the early years as in

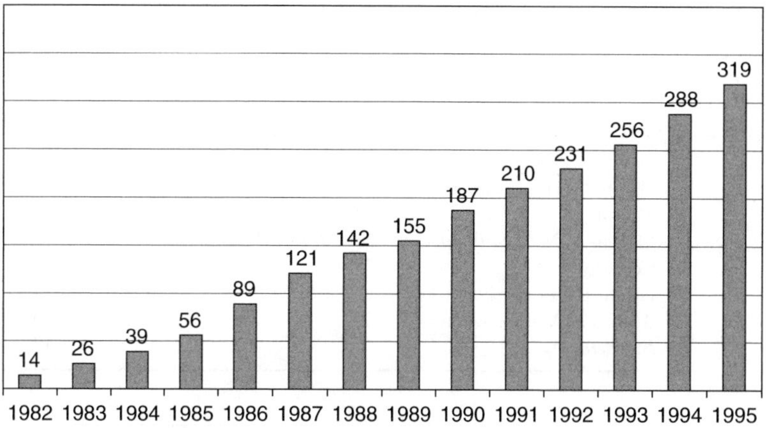

Figure 6.2 Cumulative numbers of papers published by Prusiner, 1982–95

those just preceding the awarding of the Nobel Prize to Prusiner. Overall Prusiner published over 20 papers in most years. Thus, a regular supply of prion discourse was injected in scientific literature throughout the period.

Yet, in any given year prion discourse was circulating not only via the papers published in that year but via the entire set of papers published up to that point. To get a more accurate picture of prion discourse's circulation Prusiner's publications are cumulated in Figure 6.2.

The significance of this linear tendency is easier to grasp when comparing annual and cumulative data for a given year. For instance, in 1989 Prusiner published only 13 papers (see Figure 6.1). In comparison to other years that was a bad performance. Actually, by then the scale of circulation was much more important: 155 papers had already been injected in scientific literature (see Figure 6.2).

Since our aim is to follow the circulation of prion discourse, focusing exclusively on Prusiner's numbers of papers is insufficient. A study of journals will give us clues as to the kind of circulation prion discourse enjoyed. In what follows attention is paid to the journals where Prusiner's papers were published. Table 6.3 lists the journals where he published 10 or more papers.

The journal that ranks first is a generalist and prestigious journal (mean number of two papers a year).[11] The discipline most represented is neurology (second, third and sixth ranks). This somehow puts into perspective the claim that peer-review has ceased in the TSE field

Table 6.3 Journals where Prusiner published 10 or more papers, 1982–95

Journal rank	*Journal*	*Number of papers*
1	*Proceedings of the National Academy of Sciences of the United States of America*	27
2	*Neurology*	20
3	*Journal of Neuropathology and Experimental Neurology*	18
4	*Cell*	16
5	*Federation Proceedings*	15
6	*Annals of Neurology*	14
7	*Biochemistry*	11

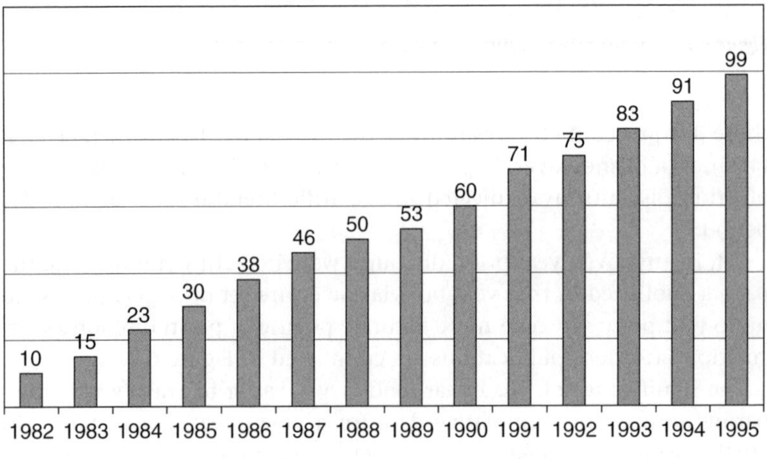

Figure 6.3 Cumulative numbers of different journals where Prusiner published, 1982–95

allegedly because Prusiner targets journals in related fields in order to avoid having his papers reviewed by his competitors (Taubes, 1986). If a number of Prusiner's publications addressed the neurology community, note however that biochemistry also features at the seventh rank.

Since over a third of Prusiner's papers were published in only 7 journals, this may suggest a tightly knitted prion chain. This conclusion does not hold if we look at Figure 6.3.

Figure 6.3 shows that Prusiner's 319 papers were published in as many as 99 different journals. Though not indicated in the figure, these journals addressed a range of audiences from ophthalmologists to protein

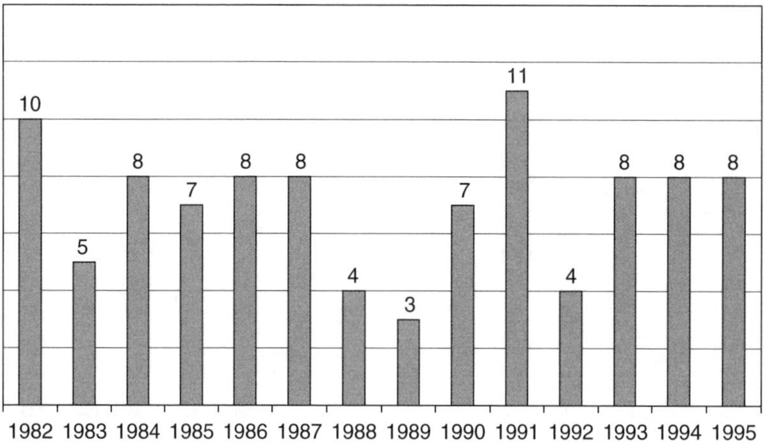

Figure 6.4 Annual numbers of journals where Prusiner published for the first time, 1982–95

engineers, through veterinarians and photochemists. The pool of journals likely to publish Prusiner's work was not unlimited and we would expect the numbers of new journals to increase in the first years and to reach a plateau at some point. Instead we find a linear progression, suggesting that he regularly published in new journals over the entire period.

Figure 6.4 shows that Prusiner published on average in 7 new journals every year. Importantly, it confirms that the tendency was sustained over the period. Note for instance that the numbers are the same in 1995 and in 1984.

Even though the numbers of new journals are not decreasing, there still might have been a relative decline if the total number of journals publishing Prusiner was increasing over the period. For instance, had he published in twice as many journals in 1995 as in 1984, the same absolute number of new journals in each year (8) would nevertheless indicate that new territories for the spread of prion discourse were running down.

The only way to determine the significance of the constant number of new journals year after year, is to look at the proportions of new journals out of the total numbers of journals that published Prusiner over the period. These are given in Figure 6.5.

By 1995 Prusiner had already published in the rather high number of 99 different journals. Yet 40 per cent of the journals where he published

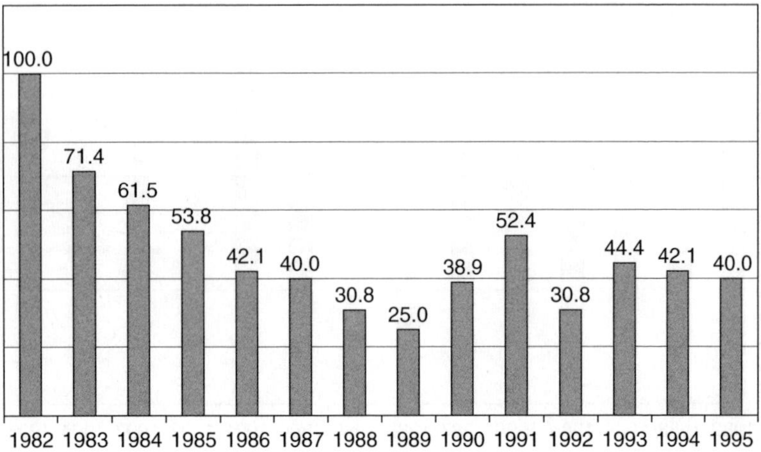

Figure 6.5 Proportions of new journals where Prusiner published, 1982–95 (%)

that year were journals that had never published his work before. From 1986 to 1995 the proportions of new journals remained constant at nearly 40 per cent. This is remarkable indeed. Figure 6.5 demonstrates that far from circulating in closed circles prion discourse constantly opened up new territories. We may speculate that several factors have contributed to its ability to reach very substantial numbers of new audiences. One was certainly Prusiner's capacity to establish collaboration with several laboratories and scientists worldwide.

Study of the journals where Prusiner's papers were published can give us an approximation of the numbers of people exposed to prion discourse during this period. Leaving aside referees and assuming that every journal had an editorial board of 5 members with no overlap, this would mean that nearly 500 key people in many different disciplines had been exposed to prion discourse up to the March 1996 announcement. We will see with the analysis of Prusiner's citations that this number is in fact a gross underestimation.

Prusiner's citations[12]

Prusiner's citations are indicators of the circulation of prion discourse in three different ways. Firstly, they constitute evidence that citers were exposed to prion discourse. Two remarks are necessary here. On the one hand, exposure by no means presupposes cognitive 'influence'. In this respect there is no difference whatsoever between positive citations and

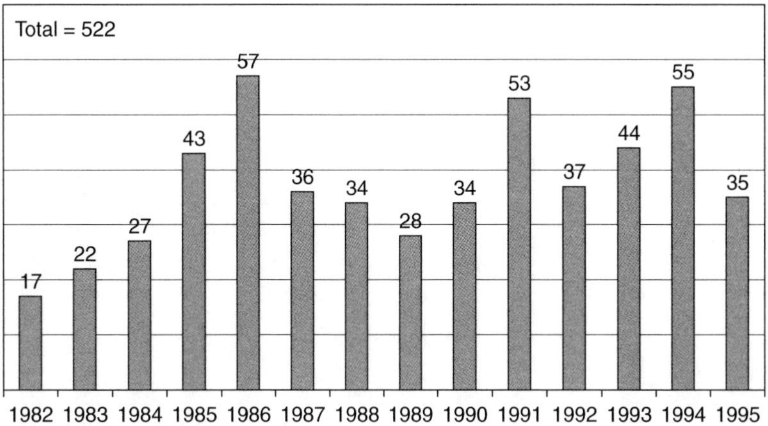

Figure 6.6 Numbers of citations received by 1982 paper, 1982–95 (including self-citations)

perfunctory and negational citations.[13] All indicate circulation. On the other hand, exposure does not even presuppose that citers actually read the papers they cited. As mentioned above, Prusiner's titles make abundant use of prion terminology and this is enough to speak of exposure of citers to prion discourse. Secondly, citations indicate exposure of citers' referees, editorial boards and readers. This is obviously the case when citers' discourse itself belongs to prion discourse but, again, exposure could involve solely those Prusiner's titles put in citers' reference sections. Finally, it is worth mentioning that citations could foster the circulation of prion discourse in referring citers' readership back to Prusiner's publications, as he himself emphatically noted in his letter to Gerald Wells: 'I think if you study Diener's review and *then carefully read the original papers*, you will be embarrassed' (Prusiner, 1987: 2; my italics).

All contributors to this volume refer to Prusiner's famous 1982 *Science* review paper in which he coined the term 'prion' (Prusiner, 1982). This paper is indeed doubly important. From a cognitive viewpoint, Prusiner used it to suggest for the first time that the scrapie agent might consist of only protein. From a discursive perspective, this paper gave rise to prion discourse. It is therefore interesting to start the analysis of Prusiner's citations with this paper. Figure 6.6 gives the total number of citations it received over the period, along with the breakdown per year.

Data provided by the Web of Science include all citations, that is, the 522 citations received by the 1982 paper include Prusiner's self-citations.

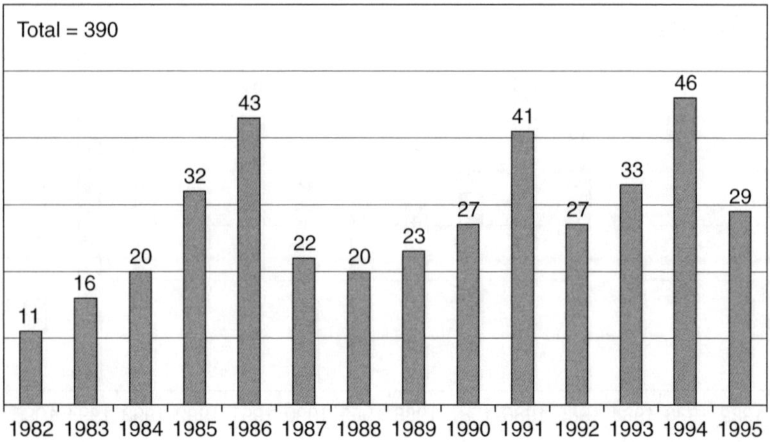

Figure 6.7 Numbers of citations received by 1982 paper, 1982–95 (excluding self-citations)

Yet, in order to assess the circulation of this article (or any other), it is obviously necessary to leave out self-citations.[14] Figure 6.7 shows that once self-citations are removed from the figures the 1982 paper was cited 390 times. Though this number looks rather high, only with a comparative analysis can we determine if it is an outstanding citation rate. Henry Small (Chief Scientist of ISI) has kindly agreed to make a special calculation for me. A 1982 paper published in *Science* is expected to have been cited 92 times by the year 2000. This confirms that with 390 citations up until 1995, Prusiner's paper was massively cited.

There is another way to assess the significance of this citation rate. Small has also done a special computer run to get data on the numbers of citations received by scientific papers 10 years after their publication. His findings are summarized in Table 6.4. These data confirm that the 255 citations Prusiner's paper had received by 1991 represent an outstanding citation rate.

We can now take a wider focus and look at the citations received by Prusiner's full publication list. Figure 6.8 gives the numbers of citations Prusiner received, along with the mean number of citations by citing paper. Between 1982 and 1995 Prusiner received 11,834 citations, with a mean number of 4.4 citations by citing paper. When self-citations are removed, the figures are as follows: 7422 citations, with a mean number of 3 citations by paper. That is a drop of 30 per cent in the mean

Table 6.4 Mean numbers of citations received by 10-year-old papers

Average scientific paper	14
Biochemistry paper	22
Neuroscience paper	24.7
Molecular biology paper	38.6
Prusiner 1982 review paper	255*

*1982–91; self-citations excluded.

number of citations, due to the fact that Prusiner's self-citation rate is very high indeed. From 1982 to 1995 he cited himself on average 19.2 times in every paper he published.

As we have seen above, critics have made it clear that the use of bibliometric methods is full of pitfalls and should always be thought through. For instance, to assess the impact of papers it has been proposed to use the number of citing authors rather than the number of citations (Dieks and Chang, 1976).[15] Similarly, we have seen that for the study of discursive circulation a necessary methodological precaution is to leave out self-citations. Here another crucial distinction to make is that between citations and citing papers. When looking at the figures for a single paper, they are equivalent: if a paper has received 5 citations it has been cited by 5 different papers. However, when looking at the figures for a set of papers such as Prusiner's publication list, citations and citing papers are no longer the same. This is because a paper can cite several publications by the same author. This was the pattern displayed in Figure 6.8: Prusiner's citers cited him on average three times in each of their papers. The distinction between citations and citing papers is illustrated in Figure 6.9. Paper 2 cites both papers a and b. Therefore the total number of citations received by a and b combined is 8, whereas the number of citing papers is 7.

For the analysis of discursive circulation whether a paper cites one or several articles by Prusiner is immaterial. In either case there has been exposure of the citer to prion discourse, and exposure of the citer's referees, editorial boards and readers. As far as discursive circulation is concerned, the number of citations is therefore redundant. More importantly, citations do not allow us to quantify discursive circulation. This is because, as we have seen above, the prion chain is the set of texts that bear traces of prion discourse. Hence, in scientific literature the units that make up the chain are not the citations but those papers that

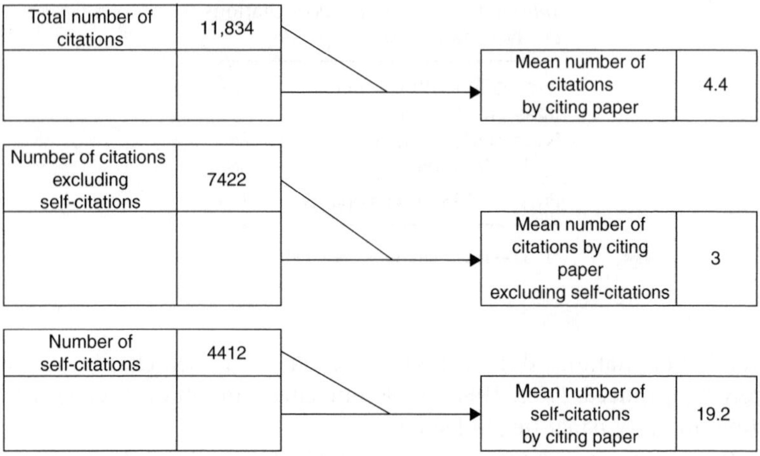

Figure 6.8 Partial overview of Prusiner's citations, 1982–95

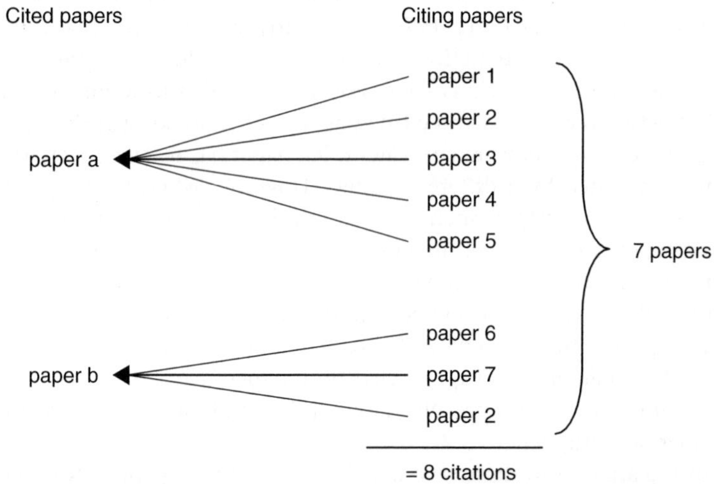

Figure 6.9 Difference between citations and citing papers

contain them. In what follows our concern will be with the papers that cited Prusiner between 1982 and 1995.

Drawing upon the distinction between citations and citing papers, Figure 6.10 gives the data that were missing from Figure 6.8 and distinguishes Prusiner's citations and citing papers.[16] The 7422 citations

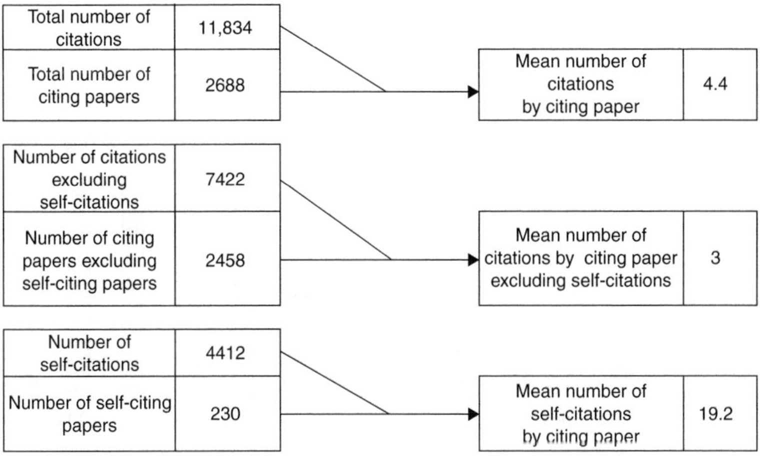

Figure 6.10 Full overview of Prusiner's citations, 1982–95

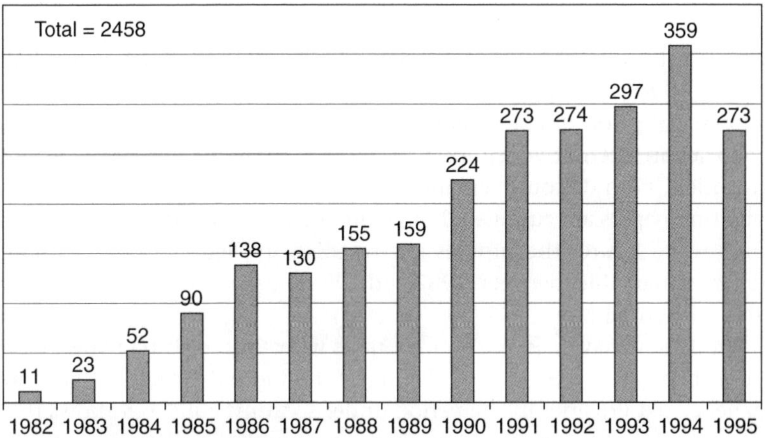

Figure 6.11 Annual numbers of papers citing Prusiner, 1982–95 (excluding self-citing papers)

Prusiner received were made in 2458 different papers. In so far as our aim is to quantify the prion chain this latter figure is the important one. Figure 6.11 gives the distribution of these citing papers in time. Note that until 1986 the progression in the numbers of citing papers was exponential. This demonstrates that prion discourse immediately and massively circulated in scientific literature. After 1986, this tendency

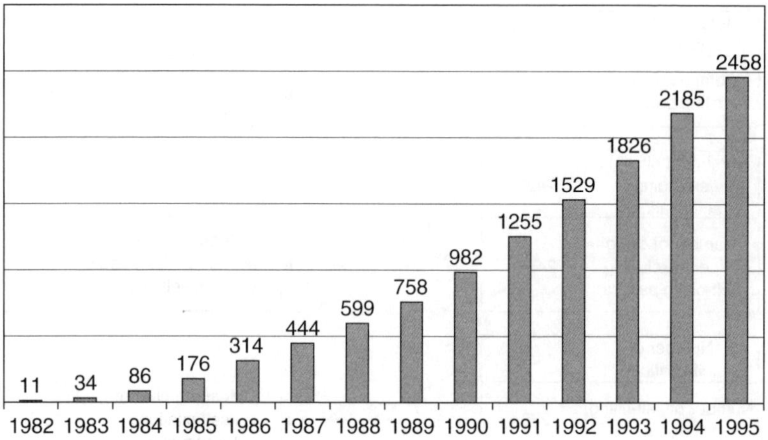

Figure 6.12 Cumulative numbers of papers citing Prusiner, 1982–95 (excluding self-citing papers)

ended but every year over 100 papers cited Prusiner. And in the first half of the 1990s, the figures were close to 300 citing papers a year. This shows that prion discourse enjoyed very substantial circulation.

To get an alternative and perhaps more accurate picture of the extent to which prion discourse circulated in scientific literature, the numbers of citing papers are cumulated in Figure 6.12. If 2458 citing papers is an impressive figure, the progression revealed in Figure 6.12 is without doubt remarkable: between 1982 and 1995 the prion chain expanded at an exponential rate.

Let us now have a look at the journals where these citing papers were published. Table 6.5 gives the listing of the journals where 20 or more papers cited Prusiner in the period under scrutiny. Table 6.5 shows that the one discipline massively reached by prion discourse was unsurprisingly neurology (13 journals). In *Acta Neuropathologica* alone, on average 5.71 papers cited Prusiner every year. Yet, two other important disciplines were reached: virology (third rank) and chemistry (fifth and seventh ranks). Scientists from other fields were also exposed via leading journals such as *Lancet* (mean number of 3.07 citing papers a year) and via very prestigious generalist journals like *Nature* (mean number of 3.64 citing papers a year).

If an important proportion were published in the 27 journals in Table 6.5, we should not jump to the conclusion that citing papers were

Table 6.5 Journals where 20 or more papers cited Prusiner, 1982–95

Journal rank	*Journal*	*Number of papers*
1	*Acta Neuropathologica*	80
2	*Proceedings of the National Academy of Sciences of United States of America*	73
3	*Journal of General Virology*	70
4	*Neurology*	57
5	*Journal of Biological Chemistry*	54
6	*Nature*	51
7	*Journal of Neurochemistry*	48
8	*Annals of the New York Academy of Sciences*	44
9	*Lancet*	43
10	*Journal of Neuroscience Research*	41
11	*Annals of Neurology*	39
12	*Brain Research*	38
13	*Neuro Science Letters*	38
14	*American Journal of Pathology*	36
15	*Journal of Virology*	34
16	*Neurobiology of Aging*	31
17	*Neuropathology and Applied Neurobiology*	26
18	*Science*	26
19	*Biochemical and Biophysical Research Communications*	25
20	*Laboratory Investigation*	24
21	*Medical Hypotheses*	24
22	*Cell*	23
23	*Journal of the Neurological Sciences*	23
24	*Journal of Neuropathology and Experimental Neurology*	22
25	*Molecular Brain Research*	22
26	*Neuron*	22
27	*New England Journal of Medicine*	21

concentrated in a relatively small number of journals. Figure 6.13 gives the numbers of different citing journals over the period. The 2458 papers that cited Prusiner were published in the no less impressive number of 528 different journals. The linear tendency of the curve suggests that he was regularly cited in new journals, which is confirmed in Figure 6.14. Note that as late as 1994 Prusiner was cited in the unprecedented number of 60 new journals. Throughout the period he was cited in dozens of new journals every year, and this tendency was clearly sustained. However, there may have been a hidden decrease if the total number of citing journals increased over the period.

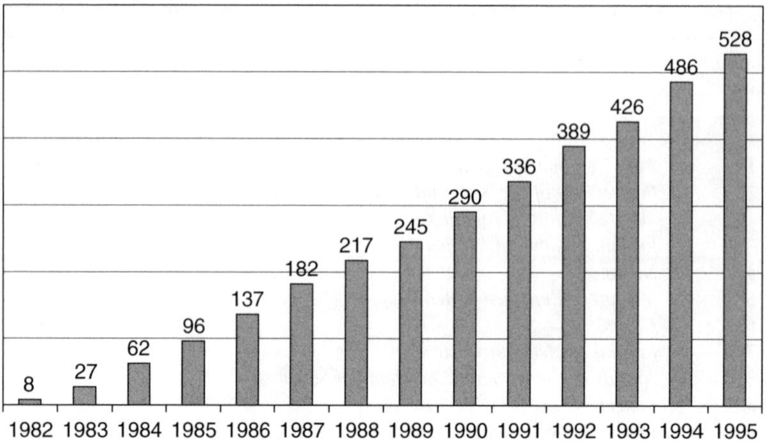

Figure 6.13 Cumulative numbers of different journals that cited Prusiner, 1982–95

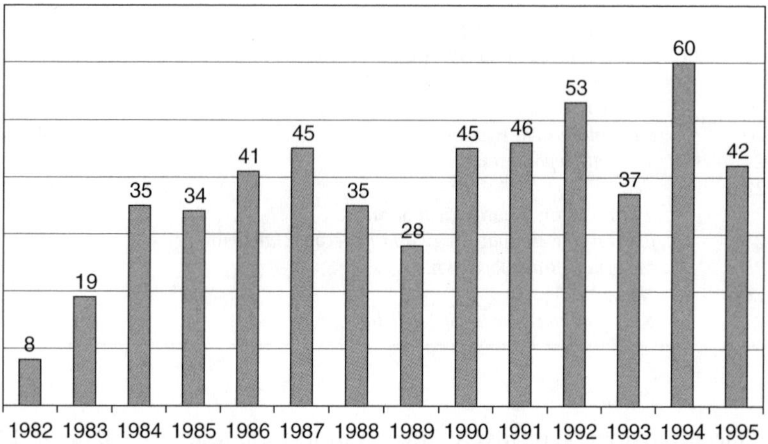

Figure 6.14 Annual numbers of journals that cited Prusiner for the first time, 1982–95

As we have done for Prusiner's publications, the only way to adequately assess the significance of the constant number of new citing journals, is to study the proportions of new journals out of the total number of journals. These proportions are shown in Figure 6.15. Again, given that the pool of journals likely to cite Prusiner was not unlimited we would expect the proportions of new journals to be declining. On

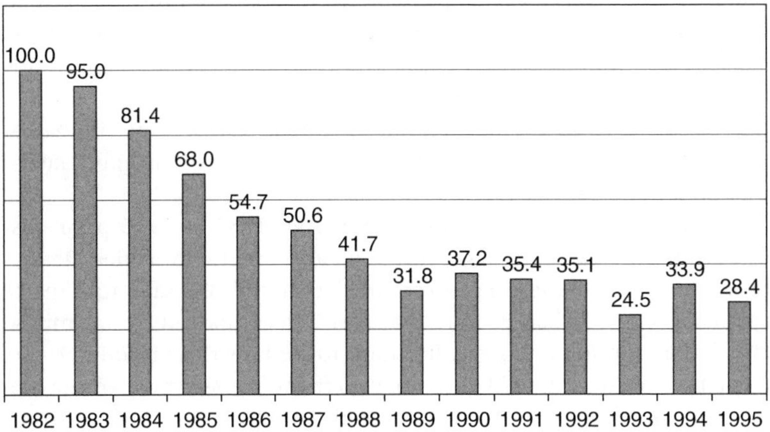

Figure 6.15 Proportions of new journals that cited Prusiner, 1982–95 (%)

the contrary, from 1989 to 1995 roughly *a third* of the journals that cited him had never done so before. This is absolutely remarkable. Thus, Figure 6.15 confirms the observation derived from Figure 6.5: far from circulating in closed circles prion discourse enjoyed highly diversified circulation. Figure 6.15 also extends this observation: if the diversified outlets of Prusiner's publications may have resulted from Prusiner's deliberate strategy, the same cannot be said of Prusiner's citations. Figure 6.15 provides undisputable evidence that irrespective of Prusiner's manoeuvres prion discourse made its way to a range of new territories.

Conclusion

The present study adds to our understanding of the prion case and its impact on the politics of BSE in drawing attention to a factor that the anthropology of science has shown to be crucial in scientific practice: numbers. Prusiner published well in excess of other TSE researchers, did so in a large pool of journals, sustained his production over the period, and constantly found new outlets for his papers. All these observations apply to papers that cited him: they were numerous and, as shown in the cumulated data in Figure 6.12, were produced exponentially, coming out in a very large number and diversified range of journals. Scientific literature was seemingly invaded by the prion discursive chain. In establishing the hegemonic status of the prion in the TSE field, this discursive circulation was probably as decisive as scientists' intellectual conviction that the prion hypothesis was accurate. Such extended discursive circulation also suggests that more than an endorsement of this hypothesis, Prusiner's Nobel Prize may have signalled the recognition of his *research programme*. Incidentally, the first sentence of the Nobel announcement defined the prion as a new 'principle of infection' rather than as an infectious 'agent'.

The present results demonstrate that from 1982 to 1995 prion discourse enjoyed early, massive, sustained and highly diversified circulation. Hence, this study has ascertained that the BSE saga took place against a scientific background that was indeed marked by an intense circulation of prion discourse. It is hard to see how this profound transformation of the TSE field could have left the management of the BSE epidemic untouched. BSE was discovered at a time when Prusiner's work was still very controversial. However, over the years the prion programme developed considerably and its discourse came to occupy a widening territory in scientific literature. If the article that announced the discovery of BSE did not bear any trace of it, there can be no doubt that later SEAC came across prion discourse while reviewing TSE literature. As March 1996 approached, SEAC and the other actors involved in the handling of the epidemic may well have been surrounded by the prion chain. Thus, the present results provide preliminary evidence that prion discourse was certainly in a position to change the perception of BSE and the course of the BSE saga.

Acknowledgements

This study would not have been possible without the expertise of information consultants, survey analysts, statisticians, computer engineers and citation analysts. I would like to thank Ola Agboola, Emmanuel Didier, Ginette Ferrié, Susan McCourt, Henry Small, Norma Williams and Ian Wilson. Special thanks to Frédéric Chauvière for proving once again that the future belongs to engineers! Finally, André Corten knows how much I owe him.

Notes

1. On the cultural significance of cannibalism in the kuru and BSE epidemics, see Seguin (2003).
2. On 22 December 1997, the British government announced the setting up of a public inquiry into the management of the BSE epidemic until March 1996. The 16-volume report of the BSE Inquiry (Phillips Report) was published in October 2000 and is available on the BSE Inquiry website (www.bseinquiry.gov.uk).

3. Results of experiments confirming that the two diseases are caused by the same agent were published in October 1997 (Bruce et al., 1997; Hill et al., 1997).
4. In 1988, the British government set up an independent expert committee to assess the risk of BSE transmission to humans. The Working Party on BSE was headed by zoologist Richard Southwood and is known as the Southwood Committee. The Southwood Report published in February 1989 remained the cornerstone of BSE management until March 1996.
5. Texts encompass minutes of meetings, scientific publications, broadcast interviews, private correspondence, journalistic articles, memos, grant applications, in-house reports, etc.
6. The Web of Science (WOS) is produced by Thomson ISI (Institute for Scientific Information). Data provided by the WOS are much more accurate than those on Medline. One reason is that the pool of journals indexed by the WOS is larger. Discrepancies between the two databases can be very important and this somehow questions the relevance of Medline for historical and sociological studies of science and medicine.
7. In bibliometrics citations made by an author are called 'references' whereas citations an author receives are called 'citations'. Thus, in what follows 'Prusiner's citations' designates the citations received by Prusiner.
8. Though ironically the Sage webpage of *Social Studies of Science* announces that 'Social Studies of Science is ranked 1 out of 26 journals in the field of *History & Philosophy of Science* in the ISI Journal Citation Reports 2002.'
9. This figure excludes contributions to edited volumes.
10. In contrast to some findings on the proportion of multi-authored papers by Nobel laureates (Inhaber and Przednowek, 1976), the present data suggest that Prusiner is a very high collaborator. Thus, if our aim was to assess the productivity of TSE scientists we should take into account that the normal count measurement favours high collaborators like Prusiner (Lindsey, 1980).
11. Traditionally *PNAS* was not a peer-reviewed journal. The route for publication was through a member of the US National Academy of Sciences. Submission was made to them and they decided whether the paper would be published. The fact remains that for many decades publishing in *PNAS* has had a very high standing.
12. See note 7 for a definition of Prusiner's citations.
13. Exponents of citation typologies have noted that certain types of citations are not indicators of influence. Perfunctory citations only indicate that some work has been performed in the same area. Negational citations indicate the citer's disagreement with the author cited.
14. Data excluding self-citations were produced with a special computer program.
15. The rationale behind this choice is that multiple citations of a paper by the same authors do not provide any new information. Consequently, the paper's impact can only be revealed by the number of different authors who cited it.
16. The numbers of citing papers were produced with a special computer program.

References

Bruce, M. et al. (1997) 'Transmissions to Mice Indicate that New Variant CJD is Caused by the BSE Agent', *Nature* 389: 498–501.

Chubin, D. and S. Moitra (1975) 'Content Analysis of References: Adjunct or Alternative to Citation Counting?', *Social Studies of Science* 5: 423–41.
Collins, H. (1974) 'The TEA Set: Tacit Knowledge and Scientific Networks', *Science Studies* iv: 165–86.
Dieks, D. and H. Chang (1976) 'Differences in Impact of Scientific Publications: Some Indices Derived from a Citation Analysis', *Social Studies of Science* 6: 247–67.
Edge, D. (1979) 'Quantitative Measures of Communication in Science: a Critical Review', *History of Science* xvii: 102–34.
Faye, J. P. (1972) *Langages totalitaires. Critique de la raison/l'économie narrative.* Paris: Hermann.
Frank Fox, M. (1983) 'Publication Productivity among Scientists: a Critical Review', *Social Studies of Science* 13: 285–305.
Fuller, S. (2000) *The Governance of Science*. Buckingham: Open University Press.
Garfield, E. (1970) 'Citation Indexing for Studying Science', *Nature* 227: 669–71.
Gilbert, N. and M. Mulkay (1984) *Opening Pandora's Box: a Sociological Analysis of Scientists' Discourse*. Cambridge: Cambridge University Press.
Godin, B. (2002) 'Outline for a History of Science Measurement', *Science, Technology, and Human Values* 27(1): 3–27.
Hill, A. et al. (1997) 'The Same Prion Strain Causes vCJD and BSE', *Nature* 389: 448–50.
Inhaber, H. and K. Przednowek (1976) 'Quality of Research and the Nobel Prizes', *Social Studies of Science* 6: 33–50.
Lindsey, D. (1980) 'Production and Citation Measures in the Sociology of Science: the Problem of Multiple Authorship', *Social Studies of Science* 10: 145–62.
MacRoberts, M. and B. MacRoberts (1986) 'Quantitative Measures of Communication in Science: a Study of the Formal Level', *Social Studies of Science* 16: 151–72.
Moravcsik, M. and P. Murugesan (1975) 'Some Results on the Function and Quality of Citations', *Social Studies of Science* 5: 86–92.
Pattison, J. (1998) Witness Statement No. 15, BSE Inquiry, BSE Inquiry website.
Phillips Report (2000) Volume 1. BSE Inquiry, BSE Inquiry website.
Prusiner, S. (1982) 'Novel Proteinaceous Infectious Particles Cause Scrapie', *Science* 216: 136–44.
Prusiner, S. (1987) Letter to Gerald Wells, 15 December, 1–2, BSE Inquiry website.
SEAC (1990) Opinion on the Public Health Implications of Eating Beef and the Epidemic of BSE, 24 July, Letter: 1–2, Annex: 1–10, BSE Inquiry website.
SEAC (1996) Statement of 20 March, CJDSU website.
SEAC (1998) Witness Statement No. 14, BSE Inquiry, BSE Inquiry website.
Seguin, E. (2000) 'The UK BSE Crisis: Strengths and Weaknesses of Existing Conceptual Approaches', *Science and Public Policy* 27(4): 293–301.
Seguin, E. (2001) 'Narration and Legitimation. The Case of In Vitro Fertilisation', *Discourse & Society* 12(2): 221–41.
Seguin, E. (2002) 'L'évaluation britannique du risque de transmission de la maladie de la vache folle aux humains', *Revue française d'administration publique* 103: 399–409.
Seguin, E. (2003) 'The BSE Saga: a Cannibalistic Tale', *Science as Culture* 12(1): 3–22.
Seguin, E. (2004) 'The BSE Crisis of March 1996 as a Discursive Event', forthcoming.

Taubes, G. (1986) 'The Game of the Name is Fame. But Is it Science?', *Discover*, December: 28–52.
Wells, G. (1988) Letter to Stanley Prusiner, 18 February, 1–2, BSE Inquiry website.
Wells, G. et al. (1987) 'A Novel Progressive Spongiform Encephalopathy in Cattle', *Veterinary Record* 121: 419–20.
Westaway, D. and S. Prusiner (1990) 'Link between Scrapie and BSE?', *Nature* 346: 113.
Will, R. (1989) CJDSU Grant Application, 6 December, 1–21, BSE Inquiry website.
Working Party on BSE (1989) *Report*. DH/MAFF (Southwood Report).

Conclusion: the Prion Case

Eve Seguin

Since 1996 prions have become part of our language and cultural landscape, like microbes and other familiar if hidden causes. But for a long time prions had an air of scandal attached to them. They allegedly threatened established knowledge, especially that on the mechanisms of infection. The notion that the prion was a foreign protein that might self-replicate in the absence of nucleic acid, as postulated in Stanley Prusiner's initial version of the prion hypothesis, was variously met with incredulity, anger, amusement or curiosity. In 1985, the discovery that the prion protein is manufactured by a host gene and exists in both normal and diseased brain might have led him to abandon his ideas or even to give up TSE research. Indeed, such a discovery may well have sounded the death knell for the hypothetical prions. But instead, Prusiner put forward that the disease-associated isoform of the prion protein is the infectious agent of TSEs as it converts the host normal prion protein into disease-associated molecules. He developed an ambitious research programme and worked on it for the next 12 years, joined in this task by a growing number of scientists. In 1997, prion research was formally recognized by the award to Prusiner of the Nobel Prize for medicine.

One might assume from this award that the nature, mode of action and pathogenecity of prions have been successfully negotiated by scientists and are now agreed. In other words, Prusiner's Nobel Prize would signal a consensus in the scientific community. We have seen that this is not the case. Every chapter of this volume is, in part, an attempt to explain how and why a seemingly stabilized scientific theory turns out to be disputed by several scientists. If there is, in fact, no consensus among TSE researchers, why is it that prions are so widely depicted as the infectious agents of TSEs? This volume has sought to elucidate some of the factors that came into play in the establishment of the hegemony of the prion theory.

Poulsen and Andersen have shown that the history of TSE research is marked by an accumulation of research anomalies and by various attempts to explain the infectious agent in unconventional terms. That Prusiner proved more persuasive than his predecessors in promoting the protein-only theory was due to his training and professional background, the conformity of his approach with the disciplinary norms of biomedicine, and his constant follow-up of suggestions by experimental work.

Kim has shown that the prion hypothesis and the empirical evidence that was produced in support of it were not definitive for the exponents of the virino hypothesis or for other prion sceptics. The controversy between the two camps was due to a clash between two different styles of research. That of Prusiner and colleagues was in keeping with current developments in biomedicine such as molecularization, standardization and commercialization, and this eventually gave the prion hypothesis more credibility within the scientific community.

Dressel has shown that the development of the prion hypothesis in Germany displays interesting similarities with Kuhn's model of scientific revolutions. It was relentlessly pushed by a charismatic scientist – Prusiner, who was followed by many young researchers whose backgrounds were not in virology. A prion research programme has been developed which explains some aspects of the TSE agent more satisfactorily than the classical paradigm of 'infectiology'. A new paradigm is under construction in TSE research, though the old one is far from dead.

Segal and Francoeur have shown that visualization was a key element in the development of prion work. The prion hypothesis holds that TSEs are caused by the misfolding of the prion protein. Thus, over the years, the race to show the tertiary structure of this protein and the production of computer graphics became centre-stage in the work of Prusiner and other researchers. Visualization of the prion protein structure considerably hardened the status of prions and helped turn the prion hypothesis into a model, as is especially clear in contemporary work on yeast.

Chamak has shown how prion research and the prion hypothesis could be stimulated by political factors. The BSE crisis in France forced the authorities to implement a range of measures, including substantial research funding, which led a number of new and young researchers into the new prion research programme. Media coverage of spongiform encephalopathies also increased, and the media's boundary work contributed to the apparent resolution of the prion controversy in France.

Seguin has shown that prion discourse, no matter how heretical it seemed, enjoyed early, massive, sustained and highly diversified circulation. Between 1982 and 1995 Prusiner published a very large number of papers in a wide range of journals which were massively cited. Prion discourse thus invaded scientific literature and reached many scientific communities. In addition to being central to the hegemony of prion research in the TSE field, such discursive circulation may well have impacted on the management of the BSE crisis.

In the last 30 years or so, historians and sociologists have endeavoured to go further than traditional histories of scientific development. For example, they have tried to understand how and why theories have been promoted, without applying later judgements as to their truth. It is now widely acknowledged that pragmatic factors play a prominent role in the acceptance or rejection of scientific theories.

Readers should therefore keep in mind, as they scrutinize the prion case, that this volume is, in part, an acknowledgement of the social competence that Stanley Prusiner and his colleagues put to work in promoting the prion hypothesis and in developing the associated research programme.

Index